Evolving Animals

The Story of Our Kingdom

What do we know about animal evolution in the early twenty-first century? How much more do we know today than Darwin did? What are the most exciting discoveries that have been made in the last few decades?

Covering all the main animal groups, from jellyfish to vertebrates, this book considers all of these questions and more. Its 30 short chapters, each written in a conversational, non-technical style and accompanied by numerous original illustrations, deal equally with the pattern and the process of evolution – with both evolutionary trees and evolutionary mechanisms. They cover diverse evolutionary themes, including: the animal toolkit; natural selection; embryos and larvae; animal consciousness; fossils; human evolution; and even the possibility of animal life existing elsewhere than on Earth. This unique text will make an excellent introduction for undergraduates and others with an interest in the subject.

WALLACE ARTHUR is Emeritus Professor of Zoology at the National University of Ireland, Galway. He is one of the founders of the interdisciplinary field of evolutionary developmental biology (evo-devo), and has a special interest in explaining scientific concepts in plain, non-technical language. He is the author of nine previous books, including *Biased Embryos and Evolution* (Cambridge, 2004) and *The Origin of Animal Body Plans* (Cambridge, 1997).

Evolving Animals

The Story of Our Kingdom

WALLACE ARTHUR

Emeritus Professor of Zoology
National University of Ireland, Galway

Illustrations by
STEPHEN ARTHUR

CAMBRIDGE
UNIVERSITY PRESS

CAMBRIDGE
UNIVERSITY PRESS

University Printing House, Cambridge CB2 8BS, United Kingdom

Cambridge University Press is part of the University of Cambridge.

It furthers the University's mission by disseminating knowledge in the pursuit of education, learning and research at the highest international levels of excellence.

www.cambridge.org
Information on this title: www.cambridge.org/9781107049635

First published 2014

Printed in the United Kingdom by Clays, St Ives plc

A catalogue record for this publication is available from the British Library

Library of Congress Cataloguing in Publication data
Arthur, Wallace.
Evolving animals : the story of our kingdom / Wallace Arthur; illustrations by Stephen Arthur.
    pages cm
Includes bibliographical references and index.
ISBN 978-1-107-62795-6 (Paperback) – ISBN 978-1-107-04963-5 (Hardback)
1. Evolution (Biology) I. Title.
QH366.2.A775 2014
591.3'8–dc23   2014007080

ISBN 978-1-107-04963-5 Hardback
ISBN 978-1-107-62795-6 Paperback

In memory of two inspirational mentors in the field of animal evolution

Alec Panchen (1930–2013)
vertebrate palaeontologist

Bryan Clarke (1932–2014)
population geneticist

Contents

Preface

When our planet was only half its current age it was already teeming with life, yet not a single animal swam in its oceans, walked on its land, or flew in its skies. Now, in contrast, there are well over a million known species of animals on Earth. Sometime in between, the very first animal arose from a unicellular ancestor. This animal was probably a tiny marine creature whose body consisted of just a handful of cells. One way of looking at the animal kingdom is as a vast number of lines of descent – or lineages – radiating out through time from that original animal, with each lineage either terminating in an extinction or continuing to evolve today.

Each line of descent has its own story to tell. So the story of the animal kingdom is a composite one, with many subplots being played out in individual lineages. In between a single lineage and our whole kingdom lie the stories of particular animal groups. In this book, I try to tell some of the individual stories, notably the human one, and some of the group stories, for example those of the three biggest groups of animals (the arthropods, the molluscs and the vertebrates). From these accounts, the composite story of the animal kingdom gradually emerges.

Often, biologists distinguish between the *pattern* and the *process* of evolution. The former concerns relationships – the issue of which types of animal are most closely related to which other ones. The latter concerns the mechanisms by which evolution comes about, including Darwinian natural selection. There have been major advances in both areas in the last two or three decades, with the result that our current view of evolution is considerably different from the view that prevailed in the middle of the twentieth century. In terms of patterns of animal relationships, a radical reappraisal of our perception of these began in the 1990s; and our views have been refined ever since through the use of DNA data to build more accurate evolutionary trees. In terms of process, the comparative study of embryonic development using modern techniques has yielded new insights into the way in which evolution works at the level of the individual animal. These insights complement earlier ones concerning how evolution works at the level of population and species. Ultimately, our theory of evolution must incorporate insights into the key mechanisms operating at both of these levels.

Not only can a story have many facets, but it can be told in many ways, each appropriate for different kinds of reader. This book is intended for anyone with an interest in the animal kingdom, its history, and how it came to be as we find it and not otherwise. All the chapters are short and are written in a conversational, non-technical way. This means, I hope, that the book will appeal to the general reader, as well as to students of zoology and other biological sciences. Also, the structure of the book is designed to ensure variety in the sequence of topics encountered, with chapters about evolutionary pattern interspersed with chapters about process.

The pictures are very much part of the story. All of the illustrations and diagrams herein are original and were commissioned specifically for this book. They are varied in type, including many evolutionary trees and several pictures relating to animal development. However, there are quite a lot of illustrations that are simply pictures of animals. It's important for readers to be able to picture in their minds animals that are not familiar to them. These include animals that are very small (for example, millimetre-long water-bears), animals that are rare (for example, fish called coelacanths), animals that are found in places we are unlikely to visit (for example, the beard-worms that are found in association with thermal vents on the seabed), and animals that are extinct and are not as well known as the dinosaurs (for example, anomalocarids and plesiosaurs). I hope that the combination of original artwork and non-technical language makes for an enjoyable read.

Acknowledgements

There are many people I would like to thank for their help in getting the book from a first draft to its present form, but with the traditional caveat that I alone am responsible for any errors of fact or failures of clarity that remain. The following people kindly read and commented upon draft material: Louise Allcock, Chris Arthur, Helen Arthur, Andy Cherrill, Ariel Chipman, Michael Coates, Erica De Milio, Uri Frank, Brian Henderson, Rose Henderson, Ronald Jenner, Grace McCormack, Dave Newton and Gerhard Schlosser. Also, special thanks should go to Robert Asher and Mark Davies, both of whom read the complete draft manuscript and provided much constructive criticism. The staff at Cambridge University Press were a pleasure to work with. I would especially like to thank my editor, Dominic Lewis, for his support throughout the period from initial proposal to final manuscript. I would also like to thank Megan Waddington, Noel Robson and Christina Sarigiannidou. My copy-editor Hugh Brazier rendered my inconsistencies consistent with a light touch, and inspired the sort of confidence an author likes to feel, but rarely does, on handing over the manuscript to a third party. Sincere thanks to everyone named above, and indeed also to the many people who are not because their work on the production of the book is invisible to the author – out of sight is not necessarily out of mind.

I   What is an animal?

What image first comes into your mind when you hear the word *animal*? A lion – 'the king of the beasts' – perhaps? Or, if you're good at mental multitasking, maybe a whole array of different creatures? The reason why one or more images will immediately flash into existence is because we all think we know what an animal is. But do we? If so, do we also have an understanding of what the animal kingdom is? Over the years, I have tried to approach these questions by doing an experiment with students, as follows. In small-group tutorials, typically taking the form of five or six students and me sitting around a table for an hour or so, asking and answering questions, an opening question I often pose, since the students are specializing in zoology, is: can you give me an example of an animal?

The reaction to this question is usually one of bewilderment. It seems too simple: is it some sort of trick? After reassurance on my part that no trick is being played, and a little clarification that I just want a common name, not a Latin one, the answers flow fast. Here is what I usually get: tiger, dolphin, elephant, cow, wolf. Of course, I don't mean that I usually get exactly those five names. So here is another example of the same kind of answer: leopard, giraffe, sheep, bat, whale.

The point I am trying to make should now be emerging. A typical response to the request to name an animal, asked of five zoology under-graduates, consists of the names of five *mammals*. It's not always the case. But it's overwhelmingly common.

There are at least three reasons for this typical kind of response. First, many students are more interested in mammals than they are in other kinds of animal. Second, my students have been mainly Irish, and the Irish countryside is scattered with domestic animals, almost all of which are mammals: hence the cows and sheep embedded in the above tutorial responses. Third, I'm convinced that the similarity of the words *animal* and *mammal* has the effect of making many people inappropriately equate them. It's only a superficial similarity, of course: the origins of the words are quite different. But it seems enough of a similarity of sound to have an effect. How many times have you heard the phrase 'birds and animals'? I've heard it far too often.

The all-mammal response to my 'name an animal' request is particularly inappropriate from the perspective of numbers of species. There are about 1.5 million named and described animal species in the world. The actual number of species is a lot higher, though it's hard to know by how much. Various biologists have contemplated this issue and have come up with guesstimated actual numbers of animal species anywhere from 3 million to 30 million. Herein I'll take the pragmatic approach of giving the approximate *known* number of species for each animal group – but it's always worth recalling that any such number is a minimum. The number of known mammal species is just over 5000. So, mammals collectively represent less than 0.5% of all animal species.

What you might call old-fashioned zoology was – and sometimes still is – taught in two parts: vertebrate and invertebrate zoology. When non-mammal species creep into my tutorial groups' responses, they are, more often than not, vertebrates. In terms of relationship with relative species numbers, an all-vertebrate response is better than an all-mammal one, but not by much. There are just over 50,000 known species of vertebrates – a composite figure including the mammals and the four other traditionally recognized vertebrate groups – birds, reptiles, amphibians and fish. So the vertebrates constitute less than 5% of animals – still just a small minority (Figure 1.1).

The old-fashioned split of the subject is itself interesting. Why divide the animal kingdom so asymmetrically for the purpose of study? I suppose the answer to that question lies in the importance of zoological studies to human and veterinary medicine; also to agriculture and aquaculture. Although the latter includes some economically important invertebrates, such as oysters, they are dwarfed, in financial terms, by fish, including farmed salmon. The former – agriculture – is almost exclusively vertebrate. It extends beyond mammals, but not much. Hens are routinely farmed, and some other birds such as ostriches are increasingly farmed too.

So, not only mammals, and not only vertebrates, are animals. All species in the animal kingdom – of which invertebrates make up the huge majority – are animals. However, that's one of those vacuous circular statements that we call tautologies. If we want to *define* animals it won't do at all. But do we need a definition? Don't all biologists, and indeed many other people too, know what an animal is, and aren't they able to draw a clear line between animals and all other life-forms? Well, actually, no.

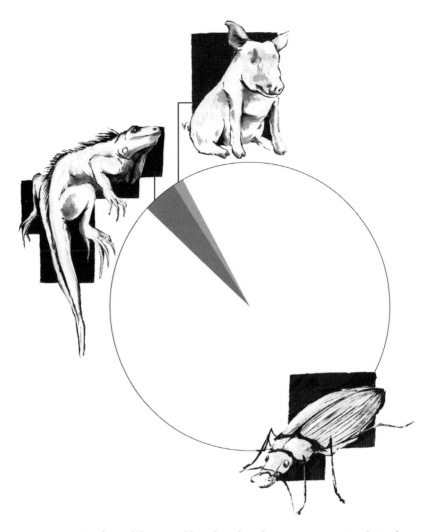

FIGURE I.I Pie-chart of the animal kingdom, based on approximate numbers of known species. The mammals, represented by a pig, make up less than 0.5% of animals. Taken together with the rest of the vertebrates, represented by an iguana, we have 50,000 or so species, but these still comprise less than 5% of animals. The other 95%-plus are invertebrates, represented here by a beetle.

Let's jump back in time to the eighteenth century, when the Swedish scientist Carl Linnaeus wrote his magnum opus, *Systema Naturae*, the first edition of which appeared in 1735. Linnaeus was attempting, in this book, to compile a hierarchical classification of life-forms – and indeed of other natural objects too, such as rocks, though his 'mineral kingdom' is no longer used.

With regard to his kingdoms of life, Linnaeus was so successful in his approach that we still use it today, albeit in modified and expanded form. His system of 'groups within groups' (a phrase often used by Darwin) extended from kingdoms all the way down to species, with a series of intermediates such as families. It is from Linnaeus that we get the system of formal Latin names for particular species, such as our own *Homo sapiens*. Many are inscrutable to those with no education in Latin, but others translate remarkably simply. Our own species name is simple enough, though whether we merit the *sapiens* is debatable. The Canadian lynx is among the most straightforward: *Lynx canadensis*.

But for now we will stick with the high end of Linnaeus's system: kingdoms. He introduced just two kingdoms of life: animals and plants. The number of kingdoms recognized has risen since Linnaeus's time, especially in the last half-century or so. It is now at least eight. Some of the extra ones are well known – such as fungi, which Linnaeus had considered to be plants – while others are not. Also, our neo-Linnaean scheme includes a category of life-forms *above* that of kingdoms – the domain. See Figure 1.2 for the relationship between domains and kingdoms.

Our domain – Eukarya, or Eukaryota – includes all those organisms that are composed of cells in which the genetic material is found in membrane-bound organelles, primarily nuclei. This type of cell (eukaryotic) differs from the simpler cells of bacteria (prokaryotic), whose genetic material is not partitioned from the rest of the cell by a membrane. The more complex structure of the eukaryotic cell renders it a better building block for making big multicellular organisms. Although some bacteria are quasi-multicellular – forming strings or mats of cells – all the truly multicellular creatures belong to the six (or so) kingdoms of the Eukarya. Here, I'll restrict attention to just one of these – Animalia, of course.

As far as we know, the animal kingdom had a single origin in the realm of eukaryotic unicells – more on this in Chapter 2. From this single origin – or stem – radiated out, during the course of evolution, the million-plus animal species of today, together with the many other animal species that have become extinct at various points in time, and so are known only from the fossils they left behind.

Now, back to the issue of defining an animal. We can see from the above that animals are multicellular (at least for most of their life-cycle)

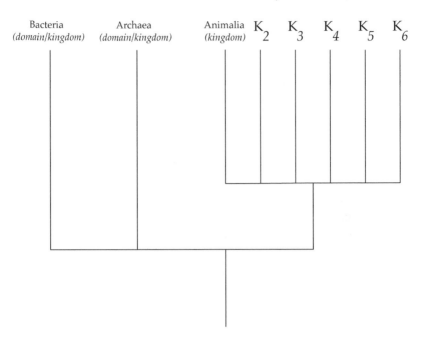

FIGURE 1.2 Domains and kingdoms of the living world, shown in simplified form. Animals belong to the domain Eukaryota (or Eukarya), which includes all those creatures made up of complex cells that have a membrane-bound nucleus and organelles. Other eukaryotic kingdoms are just labelled K2 to K6. The exact number of these is open to debate, especially as there is no clear definition of kingdom. Plants and fungi are two of them. One of the others is the kingdom containing the brown algae (most familiar seaweeds belong here; strangely, these are not plants). One name for this kingdom is Chromalveolata; an alternative name is Chromista.

and that their constituent cells are of the more complex (eukaryotic) kind. That's fine as far as it goes, and it certainly distinguishes animals from bacteria. But if we use only those two features we cannot distinguish animals from plants. So we need a third, and it's not hard to find: photosynthesis. Most plants make their own food from light and inorganic nutrients by photosynthesizing. This is something that animals cannot do.

However, nature is always messy; this makes it hard for us to catalogue it with neat categories and definitions. Not only do some plants, such as the Venus fly-trap, supplement their diet by trapping and eating insects, but others have become parasitic (usually on other plants) and have lost the ability to photosynthesize altogether. Also, some animals harbour symbiotic algae that photosynthesize. In some cases, such as certain corals, the amount of food that the animal gets in this way can constitute a considerable proportion of its total energy intake.

So far so good. We can distinguish animals from bacteria. And, with a few exceptions, we can distinguish them from plants. But our evolving definition of an animal (eukaryotic, multicellular and 'eating') would, if not refined further, include many fungi. How do we prevent it from doing so? Perhaps the obvious feature to bring in at this stage – indeed, you may wonder why I haven't brought it in already – is movement.

Generally, animals move; plants and fungi don't. But again, the effects of nature's messiness on our attempts to generalize should never be underestimated. Two of the most common invertebrates of rocky seashores are limpets and barnacles. On a casual visit to a stretch of coast we may easily observe many of each. They seem to be welded to the rock. In a sense barnacles are – they use a kind of glue or bioadhesive. But limpets are mobile – they're just rather slow and take long rests. Time-lapse photography would readily reveal this difference: over a 24-hour period, an individual limpet will move quite a bit – perhaps a few metres. But an individual barnacle will not shift its position on the rock at all.

Although *adult* plants don't move from place to place, other, non-adult stages of plant life-cycles can be very mobile indeed, as those of us who get hay fever are only too well aware. Pollen blows long distances in the wind. Also, after fertilization, seeds can move long distances too. And if we turn to the fungal kingdom, we see that although fungi do not have pollen or seeds, they have mobile spores.

The American biologist John Tyler Bonner remarked, in 1974, that organisms do not *have* life-cycles; rather, they *are* life-cycles. Taking this enlightened four-dimensional view of animals, plants and fungi, they all move. (Even barnacles move from one bit of coast to another – this is done by mobile larvae.) So is our attempt to define animals doomed? Well, actually no: movement in animals is undertaken by life-stages, whether larval or adult, that generally are able to provide their own power, albeit they may also make use of other forces, such as water currents. In contrast, the stages in the life-cycles of plants and fungi that move are usually completely dependent on external forces, such as the wind or insects in the case of pollination.

Finally, we have a definition of animals that separates them tolerably well from organisms belonging to all the other kingdoms of life. An animal is an organism with the following characteristics: its cells are eukaryotic; it is multicellular, at least for most of its life-cycle; it obtains much of its food by eating rather than photosynthesizing; and it has at least one self-powered mobile stage in its life-cycle. To these features we might add that animal cells are not enclosed within cellulose cell-walls; this not only strengthens the separation of animals from plants, but also helps us to separate them from some strange creatures called slime moulds.

So, the use of just a few criteria can form the basis for a pragmatic definition of an animal. Not only that, but animals thus defined correspond to the evolutionary radiation of forms that we noted earlier grew from a single stem in the realm of unicells. Thus the animal kingdom is what is sometimes called a 'natural' category of life-forms: one that includes all the descendants of a particular (in this case ancient) ancestor. The technical name for this kind of group is *monophyletic*. Some long-recognized groups, notably reptiles, are *not* monophyletic groups, which is why they are no longer used 'in the trade' (more on this in Chapters 6 and 7) – though of course they are still very much used in everyday language.

Now that we know how to distinguish animals from all other life-forms, it's time to look at why our favourite organisms are given their name. Where does 'animal' come from? Its root is from the Latin *anima*, meaning soul (well, actually meaning all of the following: breath, life-giving breath, mind and soul). This raises many interesting philo-sophical questions, some of which will be dealt with later (especially in Chapter 30). But let's deal with one of them now.

A long time ago, I attended the inaugural professorial lecture of Peter Harvey, a man well known in his academic field: Buddhist studies. He had just been appointed to a chair in this field at the University of Sunderland (England, not Massachusetts) and, as tradition dictates, he was giving a formal lecture of about an hour's duration to a mixed audience of academics and students from many disciplines. Appropri-ately, he kept his lecture broad, and for the most part I and other non-Buddhism-students were able to follow it well. Among other things, he covered the idea of rebirth, or reincarnation.

At question time, I asked him how far down the evolutionary scale reincarnation would apply. For example, could a human be reborn only as a vertebrate? Or might it be possible to be reborn as a worm? If reincarnation extended into invertebrates, could a human be reborn as a sponge? I received a diplomatic and intelligent, but not very enlightening, answer: that it wasn't clear.

Not everyone realizes that sponges are animals. In western society we usually first come upon them in the bath. Although some bath-sponges are indeed the skeletons of real sponges that have died, many – especially now – are synthetic substitutes. A living sponge has flesh, albeit of a rather primitive sort, covering, and supported by, the skeleton – which is itself made of many little bits called spicules and/or a sort of tough proteinaceous matrix. The flesh of a sponge is covered in holes. Water flows in via lots of small holes and out via others or, in many cases, via a single large hole (Figure 1.3). In the intervening period, small creatures suspended in the water are 'eaten'. It seems a strange word, because sponges have no jaws or teeth; the eating consists of the sponge using enzymes to digest and assimilate its tiny prey.

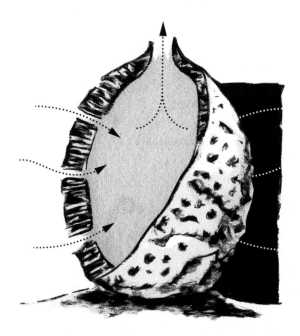

FIGURE 1.3 A sponge, shown with part of the body cut away and arrows to indicate the direction of water flow through the animal. In the example shown, water flows in through lots of small holes and out through a single large one. However, there are many variations on this theme, as might be expected for a group of more than 8000 species.

I doubt if sponges have souls. But then again do humans or any other animals have souls in the sense of spirits that transcend their bodily deaths? It's interesting that a career in evolutionary biology does not seem to narrow the spectrum of views on that question. Nor does it narrow the spectrum of views on whether there is a God (or gods). In the public at large there are theists, agnostics and atheists. Among evolutionary biologists the same three stances can be found – though it's probably true that their relative frequency is different.

The Oxford-based English biologist Richard Dawkins is the most famous evolutionary atheist. His 2006 book *The God Delusion* makes his position very clear. In contrast, we have the Cambridge-based American evolutionist Robert Asher, whose broadly theistic view is embodied in the title of his book, just as Dawkins' atheism is embodied in his. Asher's book, published in 2012, is entitled *Evolution and Belief: Confessions of a Religious Paleontologist*. In between the two 'convinced' stances – perhaps 'nearly convinced' stances would be a more accurate term – is the 'don't know' stance of agnosticism. This term was coined by Thomas Henry Huxley (1825–1895), the man who acquired the nickname 'Darwin's Bulldog' from his robust public defence of Darwin's theory of evolution by natural selection. So although there have doubtless been agnostics for centuries, Huxley was the first to wear the badge. He famously declared, in a letter written in 1886: "I am too much of a sceptic to deny the possibility of anything."

Let's return from our detour into the realm of religion to the realm of animals. As we've seen, it is possible to come up with a definition of *animal* that suffices to separate animals – more or less – from all other life-forms. Yet included within this defining umbrella lie incredibly different creatures – from sponges to people. When writing – a few paragraphs back – about my question at the end of Peter Harvey's lecture, I slipped in the phrase 'evolutionary scale' to include all animals from the most primitive, like sponges, to the most advanced, like humans. But this notion of an evolutionary scale is better avoided; it carries a heavy philosophical baggage that plagued evolutionary theory for more than a century. It has now been rightly relegated to the dustbin of scientific terms that have outlived their usefulness.

Both the origin of this term and its abandonment require some explanation. The idea of a natural scale of beings pre-dates evolutionary theory. It was used by the romantic 'nature philosophers' of the eighteenth and early nineteenth centuries as a way of ordering all the different creatures then known – a vertical, linear arrangement with humans at the top. After Darwin, it was easy to replace the romantic idea with its evolutionary equivalent. Thus, to some, evolution was seen as an escalator up which creatures moved in time, progressively getting more advanced from their primitive beginnings.

While this is true of some evolutionary lineages, including our own, it is clearly not true of others. For example, among the most successful life-forms on the planet today are bacteria. Trace any living species of bacteria back a billion years and what do you find? Essentially, a long line of ancestors all of which were bacteria-like in form. Such lineages may well outnumber those in which we might say that the creatures concerned have advanced. This is why we shouldn't think of evolution as an escalator – it can be one, but it often is not. I'll develop this point further in Chapter 11.

We've strayed out of the animal kingdom to bacteria in order to make a point. But now let's return to animals for the final point of this opening chapter. The actual animal kingdom takes the form depicted in the pie-chart shown in Figure 1.1. We already noted the asymmetry of the split between vertebrates and invertebrates. But within the latter there is another major asymmetry: insects represent about 75% of all inverte-brate animals. And within the insects, some types make up a dispropor-tionate number – beetles especially. The British biologist J. B. S. Haldane is supposed to have said, when asked by a theologian what he had learned about the Creator from his studies of the animal kingdom, "an inordinate fondness for beetles". Actually, the context in which Haldane made this remark, and its exact wording, are matters of some debate, as discussed by the American palaeontologist Stephen Jay Gould in his 1995 popular science book *Dinosaur in a Haystack* (see, in particular, Gould's chapter 29, "A special fondness for beetles").

In another of his many books, *Wonderful Life*, published in 1989, Gould asked what might happen if it were possible to 'replay the tape of life'. What he meant was the following: if we could take planet Earth back to the origin of life and let evolution happen all over again, would it take the same course? He thought not. Of course, we can't give a definite answer to this question, because we can't do the replaying experiment. One possibility would be the same broad types of animal but with different individual species. A more interesting possibility

FIGURE 1.4 Mythical animals – dragons. As with other mythical creatures (e.g. unicorns) it's interesting to contemplate why they do not exist in nature. Setting aside the difficulty of evolving fire-breathing, the top dragon would be difficult to evolve because it would require a third skeletal girdle (in between the pectoral and pelvic girdles) and a third pair of limbs. No species of reptile, bird or mammal has managed to achieve such a feat. The bottom dragon, however, is a less problematic design. Indeed, pterosaurs can be thought of as having this broad type of design, albeit different in detail.

would be different types of animal altogether. There are many fictional animals that don't exist: dragons, for example (Figure 1.4). Might these have existed if evolution had taken different courses than those that it did take?

Here, we are entering the realm of the possible versus the actual – a realm explored by the American palaeobiologist George McGhee in the concluding chapter of his 2011 book *Convergent Evolution: Limited Forms Most Beautiful*. This contrast between possible and actual can be made without invoking mythical beasts, simply by comparing the forms of existing ones. Why do insects typically have six legs while mammals never have more than four? Is it because six-legged mammals would be unfit in a Darwinian sense? Or is it that the mammalian developmental system cannot make a creature with six legs? These questions take us to the issue of what determines the directions that evolution goes in – a fascinating issue that will be explored further in later chapters.

2   Before there were animals

Although the focus of this book is on the evolution of animals, no account of such evolution would be complete without a brief look at what happened before the first-ever animal evolved from something else. What was that something? How did it evolve, and from what? How far do we have to trace evolution back until we get to the origin of life? And, before that, what processes led from an uninhabitable Earth to one on whose surface life could begin? Finally, going back even further in time, how did planet Earth come into being in the context of the evolution of the Universe?

These are big questions indeed. In science, what we perceive as the biggest questions are usually the most exciting ones, but often also the hardest to answer. Luckily, much has been learned over the last century or so in relation to some of the ancient pre-animal events referred to above – but not all of them, as we will see.

My introductory list of events in the first paragraph, above, ran backwards in time. But for the chapter as a whole we'll go forwards. We'll start fast, but we will slow down and look at things in more detail as we approach the birth of the first animal.

Cosmologists are now reasonably confident that the Universe began with the Big Bang, some 13–14 billion years ago. The Earth is just over 4.5 billion years old, so it's about a third as old as the Universe. By the way, for billions of years a common abbreviation is BY, and likewise for millions of years MY, with the addition of 'A' (BYA, MYA) if we're referring to whatever number of years 'ago'. (This is not the only system of abbreviating these vast time spans, but it's the most self-explanatory one, and thus, for our purposes here, the best; an alternative usage is Ma for millions of years and Ga for billions.)

As a zoologist, I am ill-qualified to explain the evolution of the Universe so I'll only give the briefest of accounts. If you want more detail, it's probably best to consult a popular science book written by a cosmologist – luckily there are lots of those books around.

The Big Bang, currently dated to about 13.8 BYA, is thought to have been the start of everything – not just matter and energy but space and time too. This makes little intuitive sense, but it's the prevailing current view among cosmologists. Since the Big Bang, one of the main features of the evolution of the Universe has been the formation of galaxies, with the first of these appearing quite early – about 13 BYA. Another of the main features of the Universe's evolution is expansion. This can be seen in the redshift of light reaching us from galaxies outside our own Milky Way.

This redshift deserves a brief explanation. If a galaxy is moving away from us, the light coming from it is shifted to the red end of the spectrum of visible light wavelengths. If a galaxy is moving towards us, the light is shifted towards the blue end. These correspond to shifts towards longer and shorter wavelengths respectively. These shifts are broadly equivalent to those affecting the sound waves that reach us from the siren of an emergency vehicle that is approaching (shift towards shorter wavelengths; higher pitch) or moving away (shift towards longer wavelengths; lower pitch). All galaxies show a redshift apart from our closest neighbours, where gravity can override other effects, producing a blueshift. Therefore, all non-neighbouring galaxies are moving away from our own Milky Way galaxy.

The fact that all these galaxies are moving away from us, and presumably also from each other, is evidence for the Big Bang and an expanding Universe of finite age, rather than a static Universe that is infinitely old – an idea that has now been discarded. The elimination of one of the two main competing theories of the nature of the Universe is a welcome advance in our understanding. However, in science it seems that each time a question is answered, another appears. In the case of the evolution of the Universe, it now seems that the rate of expansion has been variable over time, including a very early period of exceptionally rapid expansion that is called *inflation*, followed by a deceleration, and then renewed acceleration, of the rate of expansion. While expansion itself is potentially explicable by the explosion embodied in the Big Bang, variable-rate expansion is not – so the cause of this is still an open question.

Our own solar system is located part-way out along one of the Milky Way's spiral arms. Its formation – like the formation of other such systems – is thought to have started with unevenness in a cloud of gas and dust, leading to gravity-induced collapse of this material at a point of high density. This collapse produced our Sun, in which temperatures

FIGURE 2.1 Early Sun and proto-planetary disc. The latter contains a mixture of gas, dust and forming planets. The exact nature of planet formation varies considerably from one planet to another.

soon became high enough for nuclear fusion to begin, and a leftover proto-planetary disc of gas and dust surrounding it, from which the planets began to condense (Figure 2.1). Each planet probably grew by gradual accretion of material, coupled with more rapid growth via collisions between, and subsequent fusions of, small pre-planetary bodies. However, it's hard to generalize about planet formation: individual planets form in individual ways, as explained by the New Zealand-born scientist Stuart Ross Taylor (after whom the asteroid 5670 Rosstaylor is named) in his 2012 book *Destiny or Chance Revisited*.

The early Earth was characterized by extremely high temperatures and thus was unsuitable for life as we know it. But its gradual cooling led it to be a possible home for life by sometime around 4 BYA. The earliest fossils (of bacteria-like form) date from around 3.5 BYA. Thus the last common ancestor of all life on Earth – sometimes known as LUCA (the last universal common ancestor) – must have come into being sometime within the half-billion-year period between these two dates.

Of course, this raises the question of how we define life. The feature most commonly used to set living organisms apart from non-living things is reproduction. Rocks do not reproduce themselves; living creatures do. But there is another feature that is also important – autonomy from the surrounding environment. A rock has no real boundary – if you shatter it you end up with multiple smaller rocks. In contrast, all living organisms are bounded – usually by membranes. An exception to this is found in viruses; but there is debate as to whether these should be considered to be truly alive. A third feature of life-forms is their capacity for self-repair. While this capacity is limited – to different extents in different organisms – it is entirely absent in rocks.

Although there are different hypotheses about the exact route from non-life to life, most scientists agree that the following were stages in the process: formation of small simple organic molecules (such as amino acids); the linking up of these into large complex organic molecules (macromolecules, such as proteins); the association of multiple macromolecules into aggregates, which may have been capable of self-replication; the further refinement of these into proto-cells of greater cohesion; and the eventual attainment of a membrane-bound cell. A form of natural selection may have helped this transition to take place: at each stage, those entities with poorer powers of replication would have died out, while those with better such powers would have continued and spread. This is sometimes referred to as biochemical evolution, to distinguish it from the organismic evolution to which it gave rise.

As soon as there were cellular life-forms, Darwinian evolution – much as we know today – would have been possible. Indeed, the evolution of organisms based on the comparatively simple prokaryote cell found in today's bacteria constituted the whole of life's history for approximately its first 2 billion years. The first step towards animals came with the evolutionary origin of the more complex eukaryote cell, with its membrane-bound nucleus and organelles, at around 2 BYA. This estimate should be regarded with caution. It depends on the interpretation of various microfossils spread over a wide range of time. It should probably be considered to have an error range of plus/minus 0.5 billion years attached to it.

Although there is not yet agreement on the time of appearance of the first eukaryotic cells, there is now a consensus on the mode of their origin: engulfment of smaller cells by larger ones and a resultant symbiotic arrangement that stabilized over time, eventually leading to the smaller cells losing their individual identity and becoming organelles within the larger cell of which they became a part. Evidence for this includes the existence of small amounts of genetic material in some of the organelles (mitochondria and chloroplasts) of today's eukaryotes.

Although all complex multicellular organisms are based on the eukaryote cell, the origin of the first such cell was not rapidly followed by the evolution of multicellular forms, with complex tissues and organs. Rather, there was a long period of evolution of unicellular eukaryotes, extending for perhaps a billion years. We have already seen that there is uncertainty about the start of this period; there is also uncertainty about its end.

When did the first true multicellular creatures appear on Earth, and what were they? The first fossils of multicellular eukaryotes may have been red algae. The British palaeontologist Nicholas Butterfield described fossils that appear to belong to this group dating from as long as 1200 MYA, in a paper published in 2000. However, the first animal-like multicellular creatures are not found until much later. They derive from a phase of Earth history called the Ediacaran period, which extended for almost 100 million years, from about 635 to about 542 MYA. As we will see later, the Cambrian period, perhaps the most exciting geological period of all for animal fossils, followed the Ediacaran one – i.e. it started about 542 MYA. The Ediacaran period is named after the Ediacara Hills in South Australia (about 650 kilometres north of Adelaide). This is because the first fossils from this time period were found there – though subsequently others have been found from many locations (including the Charnwood Forest in Leicestershire, England), thus revealing that the 'Ediacaran biota' was in fact worldwide.

The significance of the 635 MYA start-date of the Ediacaran period is that it marks the end of a geologically brief worldwide ice age (the Marinoan glaciation). So after perhaps 15 million years of an inhospitably cold climate (this ice age began about 650 MYA), conditions for living creatures became much more favourable. Whether this was in some way the cause of the origin of the Ediacaran biota, or whether it merely facilitated the operation of some other as-yet unknown cause is far from certain, though the latter seems more likely: the first definite fossils of multicellular eukaryotes (both adults and embryos) date well into the Ediacaran period – later than 600 MYA.

The Ediacaran biota was composed of an array of life-forms that persisted from that time until the transition to the Cambrian. Their interpretation has been controversial ever since their discovery, and the controversy continues today. Some of the forms are shown in Figure 2.2. An interesting book on the Ediacaran biota is *The Garden of Ediacara* (1998), by the American palaeontologist Mark McMenamin.

FIGURE 2.2 Five types of Ediacaran creature. From top to bottom, the generic names for these fossils are *Dickinsonia*, *Charniodiscus*, *Tribrachidium*, *Kimberella* and *Spriggina*. Where these creatures fit into the tree of life is unclear. One hypothesis is that they represent members of an extinct kingdom and so were neither animals nor plants. Another hypothesis is that some of them – for example *Kimberella* and *Spriggina* – were animals, but others were not.

Were these creatures animals? If so, did they belong to groups of animals that we recognize from the Cambrian period up to the present day (such as sponges, jellyfish, molluscs, arthropods and chordates)? I suspect that the German palaeontologist Dolph Seilacher was correct when he proposed that most of them belonged to another kingdom (Vendobionta – named after the Vendian period, which is the old, superseded name for the Ediacaran) – in effect a doomed kingdom, an experiment in multicellularity that failed. But among the various enigmatic Ediacaran forms there were probably a few that did indeed belong to the more basal animal groups, perhaps including sponges and cnidarians. If this is true, indeed if any single Ediacaran creature was an animal, then the vast stretch of time I'm calling 'before there were animals' ended sometime during the Ediacaran period – let's guesstimate it as being about 600 MYA.

The largest structures composed of prokaryote cells were 'towers' of mat-like layers of material. These structures, called stromatolites, are among the earliest fossils, but living stromatolites are also found today (Figure 2.3). Although in one sense they are multicellular, in another sense they are not. Each stromatolite consists of a loose association of numerous cells with each other and with non-living matter. There are no complex connections between the cells, and nothing that could be called a tissue or an organ.

The Ediacaran creatures (whatever they were) and all other truly multicellular life-forms – including plants, animals and many fungi – were able to come into being because of the evolution of ways of sticking

FIGURE 2.3 Stromatolites, shown both whole and in section. The section reveals that the structure is made up of a series of layers, or mats. Present-day stromatolites can be quite large – up to about a metre in height.

cells together. There are several families of proteins that can exert this 'sticky' effect, including the cadherins – given that name because their adhesive capability is calcium-dependent. Thus a reasonable hypothesis would be that these and other groups of 'sticky proteins' evolved in the stem lineage of each of the major groups of multicells.

Like many reasonable hypotheses, this one turned out to be wrong. The closest relatives to the animals in the world of unicells are the choanoflagellates (Figure 2.4). The name of these aquatic unicells roughly translates as 'collar-whips': each of these cells has a 'collar' at what we can call its top; a whip-like structure called a flagellum

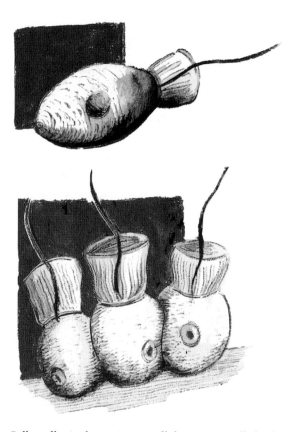

FIGURE 2.4 Collar cells. At the top is a unicellular creature called a choanoflagellate. Below is a group of collar cells, called choanocytes, as found in a sponge. Indeed, in a typical sponge this type of cell is one of the commonest in the animal's body. The similarity between choanoflagellates and the collar cells of sponges is one of the main pieces of evidence for the theory that animals arose from a unicellular ancestor that was rather like today's choanoflagellates.

protrudes from this collar and is capable of waving around, creating water currents that carry food particles.

Since each choanoflagellate is an independent organism, not part of a larger multicellular body in the way that the similar choanocyte cells of sponges are (Figure 2.4), it should have no need for cadherins. But a study of the genome of a choanoflagellate, published in 2008 by Nicole King and a wealth of co-authors, showed that a choanoflagellate does indeed have cadherins, along with the genes that make them. A reasonable question is: why? One possible answer is that choanoflagellates need to stick together occasionally for sexual reproduction to take place. Another is that cadherin proteins have two (or more) functions and that it is a non-adhesive one that is more important to choanoflagellates. In either case, the genes and proteins needed to stick the cells of a multi-cellular organism like an animal together were present in ancestral forms *before* the origin of multicellularity.

Note that sexual reproduction in effect snuck into the above para-graph with no fanfare – just as part of a hypothesis to explain something else. But this kind of reproduction itself had to evolve. It is unlikely that the replication of proto-cells in the primordial soup was of a sexual sort. Most reproduction in today's bacteria – and probably their distant ancestors of the pre-animal world – is not sexual either. Bacterial cells often simply divide in two – a process called binary fission – creating two offspring that are clones of the parental cell that they come to replace.

If studies on existing life-forms are any guide, it seems likely that sexual reproduction first became common in unicellular eukaryotes. So this is another thing – along with 'sticky proteins' – that animals (and other multicellular groups) inherited from their single-celled ancestors.

The reason why sexual reproduction is important is that it shuffles parental genes before they end up in offspring, and it thereby acts to provide variation among different individual organisms within any species – whether of a unicellular or multicellular life-form. This is crucial because Darwinian selection can only work when there is variation – and indeed specifically variation that is inherited from one generation to the next. The biochemical evolution that prevailed in the primordial soup must have acted on variation, but this was probably of a very different kind: differences in the degree of being able to reproduce at all. The organismic evolution that replaced it in the era of fully established powers of reproduction acts on variation due to mutation and to occasional (e.g. bacteria) or frequent (e.g. animals) sexual reproduction.

We'll take a more in-depth look at Darwinian natural selection later (Chapter 5). So that's enough on selection for the moment. It's now time to turn to something that is restricted to multicellular life-forms – development.

Roughly defined, development is the process of making a multicellular adult from a fertilized egg. So creatures that are unicellular throughout their life-cycles do not have development – at least not under this definition. But animals do. And so must have all multicellular eukaryotes that existed before the first animal, including the Ediacaran creatures, and also early plant-like forms such as red and green algae (Figure 2.5). Perhaps in these ancient algae, and certainly in many of today's plants, development of an adult can sometimes start not from a fertilized egg but from something else – such as a small cutting from a leaf. That's why I labelled my definition of development, above, as a rough one. Zoologists usually think of development as starting from a fertilized egg, but this is not always the case, even in the animal kingdom; for example, in many species of ants and bees, males develop from *unfertilized* eggs.

The important thing about development from an evolutionary perspective is that the evolution of multicellular creatures can only happen by deflection of the course of development from one generation to the next. There is no way that evolution can make one adult directly into another; and indeed the study of how variation in development contributes to evolution over geological time is one of the most significant themes in modern biology. That's why students of animal evolution must take development within their remit. We'll look at development itself in more detail in Chapter 13. We'll also look at its relationship with evolution in several subsequent chapters.

So let's summarize the vast period of time 'before there were animals'. It can be split into three main phases. First, the pre-Earth phase, in which the Universe appeared, followed by galaxies, including our own Milky Way, and, within these, the evolution of solar systems, again including our own. This period of pre-Earth history lasted from about 13.8 to 4.6 BYA. Second, the hot-early-Earth phase, lasting from 4.6 to around 4.0 BYA. (The geological timescale labels this, appropriately

FIGURE 2.5 Typical multicellular growth forms of red and green algae. Both of these groups belong to the plant kingdom, and indeed were among its first members. (In contrast, brown algae are now recognized to fall outside the plant kingdom.)

perhaps, as the Hadean aeon.) Third, the phase in which there were life-forms but not animal ones, stretching from about 4.0 to about 0.6 BYA, the latter date being sufficiently 'recent' for millions rather than billions to be used – so we will say that the first animal arrived sometime around 600 MYA.

Some readers may be aware of scientific studies that started in the 1990s in which the focus was on comparing DNA sequences from distantly related animals and thereby trying to come up with a date for the origin of the animal kingdom. The number of millions of years ago that the first animal, and hence the animal kingdom, was born is, according to some of these studies, more than double the figure suggested by studies on fossils: about 1200 MYA. It now seems that the early DNA studies were flawed in some way (see next chapter). More recent DNA studies have produced estimates of the age of the animal kingdom that are closer to those suggested by currently known fossils. But, as with everything in science, it is wise to be cautious. New fossils may yet turn up that will greatly modify our current understanding of when the first animal lived. So, although almost all biologists agree that there have been animals since about 600 MYA, the jury is still out on exactly how long before that the first-ever animal lived.

3   How to make a fossil

Consider a typical garden somewhere in Ireland – or, for that matter, in Britain, France or Massachusetts. Regardless of its exact size, many animals will die there on a daily basis. Most will leave no trace of their existence; within a few days or weeks it will be as if they had never lived. Far from becoming fossils for palaeontologists of the distant future to inspect and interpret, they will leave no clues as to their structure, function, or ecological context.

A good example is the death of an earthworm when a blackbird pecks it out of the top layer of the soil and eats it whole. Not only will the worm's flesh be completely digested in the alimentary canal of the bird, but earthworms have no hard parts – no teeth, shell or bones – to be egested by the bird and left on the ground for possible fossilization.

The death of an earthworm from natural causes – call it 'old age' – leads to a different scenario. The lack of hard parts means that the probability of fossilization is very low – but it is not zero. This is because, under certain unusual conditions, soft parts as well as hard ones can fossilize. The likelihood of fossilization depends on at least three things: the nature of the animal, the nature of the environment in which it dies, and the means of its death.

The animals that fossilize most readily are those with the hard parts mentioned above – shells, teeth and bones. Other animals have nearly-as-hard parts, notably insects and other arthropods. So it is not surprising that among the commonest fossils are molluscs (shells), vertebrates (teeth, bones) and arthropods (exoskeletons). Examples of each are ammonites, dinosaurs and trilobites. Animals without any hard parts, such as earthworms, leeches, jellyfish and sea anemones, hardly ever fossilize. We'll get to the reason that it's 'hardly ever' rather than 'never' shortly. But first we need to ask (and answer) the question: what *is* a fossil?

I sometimes asked this question of those tutorial groups I mentioned at the start of the book. A typical answer was 'an animal turned to stone'. Bearing in mind that the students concerned were specializing in zoology rather than palaeontology, that's not a bad answer; but it needs a little refinement.

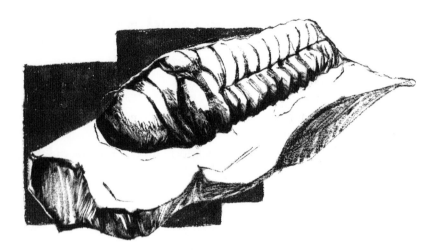

FIGURE 3.1 A fossil trilobite, with its head-end to the left. This particular fossil is still partly embedded in a piece of rock, so that only two of the three lobes that give the animal its name are visible. These fascinating animals have a substantial fossil record, largely as a result of their possession of a hard exoskeleton. Fossil trilobites span a wide range of geological time, from more than 500 MYA to about 250 MYA.

In its most inclusive sense, *fossil* refers to life-forms (including plants, fungi and bacteria – but the focus is on animals for our purposes here) that died long ago but left some evidence of their existence in today's rocks. This evidence can be a fossilized body – usually partial rather than complete – or fossilized indications of the activities of life-forms, such as burrows, footprints or coprolites (fossilized faeces). These 'preserved activities' are collectively referred to as trace fossils. In the case of animals that died recently – in geological terms – their bodies may have only partly been turned to stone; these are known as sub-fossils.

A good example of a 'classic fossil' is a trilobite – such as the one illustrated in Figure 3.1. Here, the animal has indeed been completely turned to stone. Trilobites became extinct about 250 MYA, so all trilobite fossils are at least as old as that, and many are much older – there are lots of trilobite fossils of about double that age from the Cambrian period.

Trilobites collectively constitute an extinct group of arthropods. Although we can't be certain, in life their exoskeleton was probably made from the same materials as those of present-day arthropods – such as spiders, insects and crustaceans. Of course, the exact composition of the exoskeleton varies even among these familiar current creatures.

This is apparent from the comparative rigidity versus flexibility of the exoskeletons of, say, a tarantula and a lobster. But in general a major component of arthropod exoskeletons is chitin – a carbohydrate macromolecule somewhat similar to starch.

Chitin decays more slowly than soft tissue such as muscle; thus it is more often fossilized. In general, the longer it takes for an animal – or part of it – to decay, the more likely its outline shape will be preserved during petrification – the process of turning to rock. A trilobite that died in a Cambrian ocean will have ended up on the seabed, or substratum. If this was soft – sandy, say – then the trilobite might have become buried. Later, the sand might have got compressed into sandstone; and this may still contain the trilobite's body, sometimes with the exoskeleton – and very occasionally soft tissues – partly transformed into stone. Further transformation will then occur over a longer time period – millions of years.

Sandstone is one kind of sedimentary rock. There are many others, such as mudstone and siltstone. Such rocks are formed from the compression of small particles that are themselves the product of erosion, or, as in the case of limestone, the fragmentation and crushing of biologically derived structures such as molluscan shells.

A very different type of rock, called igneous, is formed by the solidification of magma, originally from the Earth's molten mantle, following either volcanic eruption (*extrusion*) into the air or sea, or seepage (*intrusion*) into already-formed solid rocks of the Earth's crust. This type of rock – granite and basalt are examples of intrusive and extrusive igneous rocks respectively – is devoid of fossils for obvious reasons: the molten magma never contained dying animals. Any animals unlucky enough to have died by contact with the flowing lava produced from a volcanic eruption would have been incinerated and left no trace of their existence.

There is a third type of rock, called metamorphic, which is the result of a previously existing rock having been altered by extreme temperatures and/or pressures. An example is marble, which is metamorphosed limestone. Fossils are rare but not non-existent in metamorphic rocks. Those that are found have often been distorted or semi-destroyed by the same forces that have changed the rock in which they are embedded. The least affected are small, hard spores. From the perspective of the animal fossil record, metamorphic rocks make only a small contribution.

So the majority of animal fossils are of hard parts, buried in sediments and petrified in the process of rock formation. Further, the most extensive sediments, and thus the source of the most numerous animal fossils, are marine ones. Lake-beds, and even depressions in terrestrial environments, can also accumulate sediments, but the contribution these make to the fossil record overall is small compared to their marine counterparts. This is one reason why trilobite fossils are abundant while insect fossils are less so, despite the huge number of species of insects, both past and present.

I said earlier that the likelihood of an animal fossilizing depended not just on the nature of the animal (with or without hard parts) and the environment it lived in (for example marine versus terrestrial) but also on the manner of its death. I gave the example of the bird-eaten earthworm as a case of a zero probability of fossilization because the fact that it died by predation would mean that its tissues were enzymatically reduced to small molecules in the bird's stomach. My trilobite dying of disease or old age and being buried in sand provides a counter-example of a much higher probability that a fossil will form.

Normally, in the latter sort of situation, only hard parts fossilize. This is true regardless of how long ago the fossilized creature lived. It is as true for hominid fossils from the famous Lake Turkana site in sub-Saharan Africa as it is for trilobite fossils from western Canada, despite the orders-of-magnitude difference in age: fossils of proto-human species derive from the last few million years, in contrast to trilobite fossils deriving from a few *hundred* million years ago.

Occasionally, though, environmental and cause-of-death conditions are such that even some soft tissues fossilize. These rare situations are referred to by the German name *Konservat-Lagerstätten*. This roughly translates as 'well-preserved mother lodes'. There are sufficiently few of them that they are known individually by name – usually called after the relevant location. The Ediacara Hills fossil site in Australia discussed in the last chapter is one instance; the Burgess Shale of British Columbia in Canada, which will be discussed in the next chapter, is another. More recent instances include the Rhynie chert near Aberdeen, Scotland (about 400 MYA), the Mazon Creek in Illinois, USA (about 300 MYA), and the Solnhofen limestone of Germany (about 150 MYA) from which fossils of the iconic dinosaur-bird *Archaeopteryx* have been found.

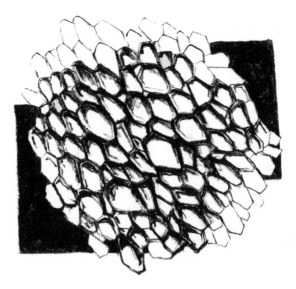

FIGURE 3.2 A fossilized piece of dinosaur skin. Although skin is generally referred to as a soft tissue, and it is certainly softer than teeth or bones, dinosaur skin was much tougher than typical mammalian skin. Also, vertebrate skin in general has an outer layer of the tough protein keratin, which doubtless renders its fossilization probability higher than that of a really soft tissue, such as the mammalian placenta.

'Well-preserved mother lodes' such as these probably often arose by rapid mass mortality and burial. The Burgess Shale fossils are thought to have resulted from the collapse of marine cliffs and the 'smothering' of the fauna below. In contrast, the Rhynie chert fossils are thought to have resulted from upwelling of water that was rich in silica from volcanic springs. It's important to note that the fossils preserved at Rhynie are terrestrial forms – including some of the earliest insects. These are two very different faunas and two very different modes of mass mortality but the result is the same: unusually good preservation, including soft tissues. However, the phrase 'soft tissues' is perhaps misleading. It includes a whole range of things: insect exoskeletons (softer than molluscan shells), muscle and skin (softer than exoskeletons; Figure 3.2 shows the fossilized skin of a dinosaur) and *really* soft tissues such as those found in embryos or jellyfish. The further along the spectrum from very hard to very soft, the rarer the preservation of the tissue concerned.

Having learned a bit about the formation of fossils, we now need to ask what is important about fossils, in the sense of what we can learn from them about animal evolution. The two most important things are establishing the group – extant or extinct – to which the fossil belongs, and finding out how old it is. The former is based on a thorough description and interpretation of all the observable features of the fossil, the latter on radiometric dating – involving measurement of the amount of 'decay' of naturally occurring radioactive isotopes – of the relevant rocks. (Note the 'relevant': more on that shortly.)

There are various methods of radiometrically dating rocks, but they all use the same principle, as follows. Any rock contains both stable isotopes of its constituent elements and unstable (radioactive) ones. The latter decay into stable forms in a highly predictable manner. So if we measure the ratio of particular unstable-to-stable element ratios, one being uranium-to-lead, we can estimate the age of the rock – the lower the fraction of the radioactive form in the rock, the older it must be. Often, these estimates have errors as low as plus/minus 1% or even less.

The starting point for measuring the radioactive decay of unstable isotopes in a rock is the point at which it has cooled enough so that there is no longer diffusion of materials between the rock and its surrounding environment. This happens as the rock forms from molten magma or lava. But note that it is implicit here that we are dealing with igneous rocks – which do not contain fossils. Thus in order to date a fossil-bearing sedimentary rock layer, it is usually necessary to perform radiometric dating on the igneous rocks immediately above and below it, or indeed those that intrude into it. So the 'relevant rocks', two paragraphs back, are igneous; and the dating of their sedimentary counterparts is indirect.

Shortly after the beginning of this chapter, we discussed the basics of 'how to make a fossil'. Then we went on to learn how to make a super-fossil, as those of the 'well-preserved mother lodes' might be called. But now we need to return to our starting point – the bird-eaten worm – and look at the opposite of how to make a super-fossil: how to avoid making a fossil at all. This time, though, we need to ask not just how individual animals might leave no fossil evidence of their existence, but how it might be that long evolutionary lineages of animals stretching through millions of years might also leave no fossil evidence. The reason why this question is interesting concerns the different estimated ages of the animal kingdom that studies of fossils and of DNA sequences produce, as briefly noted at the end of the previous chapter.

Estimating the age of the animal kingdom or of any group within it using DNA sequences involves the concept of a molecular clock. This is the idea that a DNA sequence, such as a gene, evolves at an approximately constant rate over periods of time that are long enough for short-term fluctuations to cancel out.

The way the clock technique works is as follows. First, you measure the extent of DNA sequence divergence between two animals that are reasonably closely related – say two mammals. Then you look at when their lineages diverged according to the fossil record in terms of MYA. Dividing the former by the latter gives a result in the form of 'X% difference per million years'. You can then repeat the process for more distantly related animals – say a mammal and a fish. If the molecular clock were completely regular, like a real clock – though no biologists believe that – the DNA sequence of the gene that makes a particular protein should evolve at a precisely constant rate. In this case, the division of a greater percentage difference in DNA sequence by a greater time of divergence should give the same result in terms of percentage difference per million years. Although the molecular clock is not completely regular, using the method described here generally works fine, just so long as you don't expect it to provide *precise* estimates.

This technique works well for estimating the age of most animal groups. However, it is problematic for estimating the age of the animal kingdom itself, or for estimating the ages of the earliest divergences between major animal groups. In these cases, the difference between clock-based and fossil-based estimates of the date of the relevant divergence is often large, and that leads to some difficult questions. For example, if the first vertebrate lived in the early Ediacaran period, about 600 MYA, as suggested by the molecular clock work summarized in the 2009 book *The Timetree of Life* (edited by S. B. Hedges and S. Kumar), why are there no vertebrate fossils for more than 50 million years? And if the first animals lived more than 1200 MYA, as a pioneering clock-estimate produced by the American biologist Greg Wray and his colleagues in 1996 suggested, why did animals in general leave no fossils for more than 500 million years?

There are sometimes said to be 'ghost lineages' that leave no fossil evidence of their existence, either because of a lack of readily fossilizable hard parts or for other reasons such as patchy geographical distribution. While such ghost lineages doubtless exist in many cases, two problems are encountered in appealing to this concept to explain the apparent absence of early vertebrate fossils noted above. First, why are there apparently no vertebrate fossils from the Ediacaran *Lagerstätte*, in which soft tissues fossilized? Second, what does a vertebrate with no vertebrae look like, and how does it turn into a 'real' vertebrate in all of the multiple independent lineages that must have radiated out from the initial ghost lineage? So far, these are unanswered questions – though we will return to the issue of vertebrate origins in Chapter 20.

Regarding the whole animal kingdom, and the possible existence of ghost lineages with their associated lack of fossils for more than 500 million years, the idea that so many lineages would leave no fossils at all for such a long time is hard to believe. We can see how it is possible not to make a fossil in a restricted context such as the death of a worm. But extrapolating the 'not making fossils' argument into such a broad context makes little sense. Thus while the jury is still out on the age of the animal kingdom, as noted in the last chapter, the case for the correctness of the more extreme estimates, such as 1200 million years, is looking rather weak.

One possible reason for the discrepancy between molecular and fossil-based estimates of very early divergences concerns the calibration of the molecular clock. This requires a bit of explanation. For the time from the start of the Cambrian period (and hence the start of unambiguous animal fossils) to the present, an enlightened version of the clock principle seems to work. There are some ifs and buts, though they're not too serious. Different genes evolve at different rates. And any one gene obeys what has been called a 'sloppy clock' rather than a precise one. No big deal: allowances can be made for such complications.

However, there is a specific problem that may well affect attempts to extrapolate the clock back to very ancient evolutionary history, as follows. A protein molecule evolving to do a particular job may evolve quite rapidly at first, when it is only in rough-and-ready form; it may then slow down in its evolution as it approaches the optimum form for the job concerned. Thus the calibration of the clock based on more recent evolutionary time will fail and the method will produce erroneously old estimated dates for ancient evolutionary divergences.

At the other end of the spectrum from ancient evolutionary events, such as the origin of the animal kingdom, are 'recent' evolutionary events, such as the origin of humans. We'll look at this topic in detail in Chapter 24. For the moment, I'd just like to make a few comments on 'how to make a human fossil'.

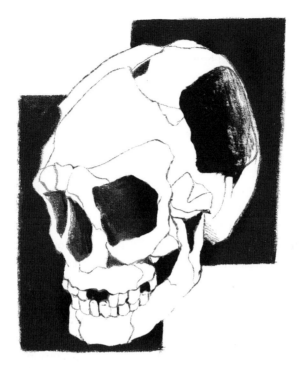

FIGURE 3.3 A fossilized human skull reconstructed from fragments. In many cases, far more fragments have to be pieced together to recreate the skull than shown here. However, in a few cases, near-complete fossil skulls have been found.

There are no human remains from true *Lagerstätten*, although our own fossil record is actually quite good thanks to thousands of conventionally preserved fossils. Thus the story of human evolution in the fossil record is told by fossilized hard parts – notably bones and teeth. Of particular interest are skulls, as they document the evolution of the large brain size that provides the basis for our humanness. Complete skulls are known, though rare. So those interested in the evolution of brain size need to painstakingly reconstruct fossil skulls, often from very fragmentary material (Figure 3.3). Then they need to estimate brain size on the basis that the brain fills the cranial cavity of the skull to the same extent as it does in modern humans.

Here we see intensely practical problems of palaeontology, as opposed to more esoteric ones such as how to connect fossil data with data from molecular-clock studies. These practical problems of piecing together bone fragments and making inferences about the soft tissues that went with them are well known among vertebrate palaeontologists. And of course equivalent problems, though usually not involving bone, are well known among invertebrate palaeontologists. The amazing thing is not that these problems exist, but that we've come so far despite them. In the following chapter, we'll see what is known about some of the fascinating animal fossils of the Cambrian period. Anyone wanting to learn more about fossils in general should read the 2009 book *Fossils: The Key to the Past*, by British palaeontologist Richard Fortey.

4   The Cambrian explosion

As we saw in Chapter 2, there have been multicellular life-forms on this planet since about 1200 million years ago (MYA). These first life-forms, however, were not animals; they were probably red algae. Later, in the Ediacaran period, which began some 635 MYA, there are fossils of multicellular creatures, some of which *may* have been animals; their interpretation remains controversial. The beginning of the Cambrian period, 542 MYA, marks the start of an abundant fossil record of creatures that undoubtedly *are* animals. In the Cambrian we are faced with a profusion of animal fossils, both those that we recognize as being clearly related to some of today's animals, and others that are more enigmatic in terms of where to place them in our 'groups within groups' system of naming and ordering animals that we inherited from Linnaeus.

Geological periods are often named after places where rocks of the relevant age are found. As we noted in Chapter 2, the Ediacaran period is named after the Ediacara Hills in Australia. The Cambrian is named after Wales. The basis of this latter naming is not as readily apparent as that of the former. But the Welsh name for Wales – Cymru – gets us a bit closer to seeing the connection. And the Roman name – Cambria – makes it crystal clear. There is even a town in Wales where the local newspaper is called *The Cambrian News*.

At this point I'd like to consider the extent to which the names of geological periods are known to those who are not scientists or natural historians. The proverbial person 'on the street' has never heard of the Ediacaran period. But what about the Cambrian? This is certainly better known – because of the 'Cambrian explosion' of animal fossils that gives this chapter its title. Even so, outside science I'd guess its familiarity is limited. The geological period that is best known outside science is the Jurassic, because of the film *Jurassic Park*. Everyone who has seen this film is aware that dinosaurs were alive during that period of Earth history.

Most of the other geological periods from which we find animal fossils do not have familiar names. Since there are 11 of them subsequent to the Cambrian, of which only the Jurassic is well known to most people, it makes sense, in a book of this kind, to refer to their names sparingly: in most cases a geological time given in the form of millions of years ago confers easier comparability with other times. I'll generally follow this formula, avoiding the names of periods (and the smaller units, such as epochs, into which periods are divided) with just a few exceptions. However, readers who are interested can find the full sequence of periods from the Cambrian onwards in the Appendix.

So, back to the Cambrian, and specifically to the 'explosion' that bears its name. Although no-one who has seriously studied the Cambrian record doubts that some sort of explosion affecting animal life occurred then, interpretations of it have differed widely. We'll look at those different views shortly. But before doing so, it's necessary to have some 'facts' – at least as far as these are possible for such an ancient event.

The Cambrian period started about 542 MYA, as noted above. It ended about 485 MYA and thus lasted some 57 million years. The explosion of animal fossils occurred during the early and middle parts of the period – extending from about 535 to 505 MYA. If we accept these figures as a rough guide to the duration of the explosion, it's immediately clear that, while it's an explosion in geological time, it's most certainly not an explosion in what is often referred to as ecological time.

For a small invertebrate with an annual life-cycle, 30 million years is 30 million generations. In terms of the time available for Darwinian selection to act, this is very long indeed. We know that natural selection acting on small invertebrates can have significant results over periods as short as 10–100 years – this is especially so in some extreme environments created by humans (e.g. fields sprayed with insecticides), and it may also be the case in extreme environments created more naturally. However, if we ask what fraction of geological history 30 million years represents, it then looks like a very thin slice of time. It's only about 5% of the history of animals, using our broad-brush 600 MYA estimate of their origin. And it's less than 1% of the time that has elapsed since the origin of life.

What these alternative ways of looking at time mean is that an event that is 'explosive' to a geologist is very slow and gradual to an ecologist. As long as we realize that fact, we can interpret 'explosion' in the right way, and not make it seem even more dramatic than it was.

We can be factual about place as well as time. There are about 10 *Konservat-Lagerstätten* dating from Cambrian times. The best known of these are from Chengjiang County in China, British Columbia in Canada, and Sirius Passet in northern Greenland. Of these, the first-discovered, and most famous, is the Canadian one, referred to as the Burgess Shale. Between them, these and the others cover a span of time from about 525 to 500 MYA.

Now, armed with information about time and space, we are in a position to consider the kinds of animals involved, though this is where it becomes hard to separate facts from opinions. One of the leading palaeontologists studying the Burgess Shale was Cambridge-based Harry Whittington, who died in 2010 at the grand old age of 94. He and his colleagues, Derek Briggs and Simon Conway Morris, suggested in a very low-key way, and in a very specialist scientific journal, that some of the animals of the Burgess Shale belonged to high-level animal groups all of whose species are now extinct.

The low-key approach to the possible nature of the Burgess Shale fossils was no longer possible after 1989, the year in which the American palaeontologist Stephen Jay Gould published his book *Wonderful Life*, with a subtitle of *The Burgess Shale and the Nature of History*. In this book, Gould strongly advocated the view that many of the Burgess Shale animals belonged to now-extinct phyla. (Phyla – such as molluscs, arthropods and chordates – represent the highest-level groups within kingdoms in the Linnaean system. The singular of phyla is phylum.) Rather than advocating by understatement, Gould advocated by 'shouting from the rooftops'.

It's probably best to look at the changing interpretations of Cambrian fossils via a few examples. I'll use three here, all of which have been interpreted in different ways by different authors.

The first of my chosen three is an animal called *Anomalocaris* (Figure 4.1) – which roughly translates as 'odd prawn' ('strange crab' and 'unusual shrimp' have also been used). As so often happens in palaeontology, fragments of this animal were discovered before nearly-complete specimens. It seems that the hardest parts, and hence the most fossilizable ones, were the mouth and the large anterior appendages. A single one of these appendages looks rather like a prawn – or at least the body of a prawn with its head missing. It was the discovery of one of these that led to the name. The (separate) discovery of a fossilized mouth

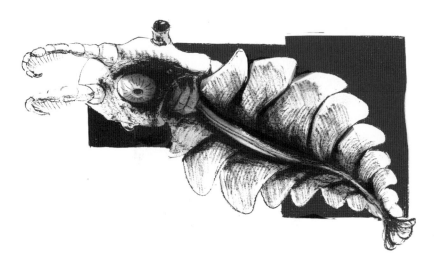

FIGURE 4.1 One well-known animal from the Cambrian period: the large marine predator *Anomalocaris*, viewed from below, as a prey animal on the sea floor might have seen it. *Anomalocaris* is thought to have swum using an undulating motion of its lateral flaps. Fossilized mouths and large anterior appendages were discovered before any nearly-complete specimens. This led to some bizarre misinterpretations: see text.

led to its identification as a jellyfish. This seems very odd at first. But the mouth is round, and, although we now know that it was composed of hard exoskeletal material, the fact that soft tissues are also preserved in *Lagerstätten* means that it *could* have been a fossilized jellyfish or other round animal. Finally, the discovery of a poorly preserved part of the body of another *Anomalocaris*, with an irregular and porous appearance, led to its identification as a sponge.

This, more than almost any other example, illustrates the dangers of trying to interpret fossil material from the Cambrian. Here we see a single animal interpreted as three, and not only that but three that belong to different animal phyla. The person who put the *Anomalocaris* jigsaw together was Harry Whittington. It was through his work that the animal shown in Figure 4.1 came to be known. These weird and magnificent beasts were among the largest animals that lived in the Cambrian seas, and are thought to have been ferocious predators. Estimates of the length of the largest specimens range up to 2 metres.

Arriving at an accurate (we hope) picture of the animal is not the end of the story. The next question is to which animal group it belonged. Of the three misidentifications of parts of it, two were hopelessly wide of

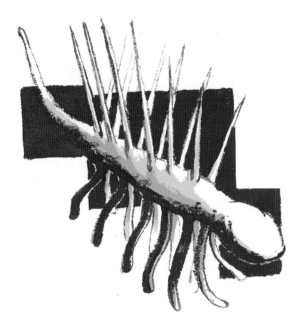

FIGURE 4.2 Another well-known Cambrian animal: *Hallucigenia*. Interpretation of this creature, like that of *Anomalocaris*, has changed over time. In particular, the original interpretation was upside down in relation to what is now thought to be the correct orientation (the one shown, with dorsal to the top).

the mark: this is no jellyfish or sponge. But the 'prawn' at least points in the right direction. Prawns are crustaceans and crustaceans belong to the phylum Arthropoda. Although Gould reckoned that *Anomalocaris*, and many other Cambrian animals, fell outside any of the phyla to which today's animals belong, there is now a consensus that it was some sort of arthropod.

That the second of my three examples is a bit weird can be seen from its name: *Hallucigenia*. It does indeed look like the stuff of dreams or drug-induced hallucinations (Figure 4.2). And yet it did exist in those same Cambrian oceans in which *Anomalocaris* swam.

I doubt if the general form of the creature that we call *Anomalocaris* will change significantly in the future. But I am much less convinced that we have pinned down the form, and the affinities, of *Hallucigenia*. The main change over time in its interpretation has been from (a) the view that it walked on its spikes and the flexible appendages on the other side of the body were tentacles to (b) the view that inverts the animal, turning the tentacles into legs (the evidence that they

FIGURE 4.3 Outline body form of an extant velvet worm. These animals are referred to as having a lobopod design: the limbs are unjointed lobes projecting downwards and outwards from the body. It is thought that the Cambrian animal *Hallucigenia* was also of this broad type of design, though with the addition of dorsal spikes.

occurred in pairs is equivocal) and the spikes into dorsal defence structures – perhaps to deter predators such as *Anomalocaris*.

Supposing there is no further significant change in our interpretation of the form of *Hallucigenia*, to what group of animals does it belong? The prevailing current view is that it's a lobopod – this is the name given to invertebrates with non-jointed walking legs. There are large (up to about 20 centimetres long) animals of this kind alive today called velvet worms, but these are restricted to the southern hemisphere, so those of us living in Eurasia or North America rarely get a chance to see them. They're regarded as belonging to a group that is closely related to, but not within, the arthropods – see Figure 4.3 for an illustration of the general body form of a velvet worm.

My third and final example of a Cambrian animal is most definitely not an arthropod. It's much closer than they are to our own – vertebrate – group. Its name is *Pikaia* and its general form is shown in Figure 4.4. Two of the key features are: a series of chevron-shaped blocks of muscle extending along the antero-posterior axis of the animal, and a stiff rod running along the dorsal midline to which the muscles were probably attached.

Up to now I have mostly been referring to 'our' group of animals as the vertebrates, a group named after their possession of a vertebral column (alias the backbone or spine). However, there are close relatives of the vertebrates alive today that lack a vertebral column. Instead, they have – as adults in one case and larvae in another (Figure 4.5) – a stiff dorsal rod like the one in *Pikaia*. This rod is called the notochord – not to be confused with our spinal cord, which is composed of nervous tissue

FIGURE 4.4 The Cambrian animal *Pikaia*, from the Burgess Shale of British Columbia, western Canada. This is thought to have been an early member of our own phylum – Chordata. However, it was not the earliest chordate: earlier representatives have been discovered among the Chengjiang fauna of China.

and is enclosed within the protective casing of our backbone. As embryos, we humans (and other vertebrates) have notochords, which function as the source of important developmental signals. But the structural role of the notochord has become redundant in the vertebrates, and it disappears before the embryo gets very far through its development.

There is a reason for this digression into three sorts of midline dorsal structures – notochords, spinal cords and backbones. Animals with backbones are, by definition, vertebrates. Likewise, animals that have a notochord are, by definition, chordates. And our phylum – recall that a phylum is the highest level of group conventionally recognized within a kingdom – is called the Chordata. Vertebrates make up just one – but by far the biggest – subphylum within this.

Ever since its discovery, *Pikaia* has been seen by most – but not all – palaeontologists as an early chordate. This is one of the few creatures on which the interpretations of Stephen Jay Gould and Simon Conway Morris are similar. You'll probably recall, from a few pages back, that

FIGURE 4.5 Two types of extant non-vertebrate chordates: a tunicate larva (left) and an amphioxus or lancelet (right). Although neither of these has a spinal column or a skull, and so they are not vertebrates, both possess a stiff dorsal rod called the notochord, which vertebrates also possess. However, in a typical vertebrate the notochord is a transient embryonic structure. In tunicates, it lasts for a greater portion of development, disappearing only when the larva metamorphoses into an adult; in an amphioxus the notochord persists throughout life.

Conway Morris was one of the palaeontologists working with Harry Whittington on the Burgess Shale animals. Although it was their work, and indeed their initial interpretations, that inspired Gould to write *Wonderful Life*, Conway Morris later became more conservative in his interpretations of many of the Burgess Shale fossils, placing them in existing animal groups rather than seeing them as falling outside these. His later interpretations, which have become generally accepted, are given in his 1998 book *The Crucible of Creation* – which is subtitled *The Burgess Shale and the Rise of Animals*.

Even if *Pikaia* is indeed a chordate, the discovery of earlier chordates, some of which may even be vertebrates, among the Chinese Chengjiang fauna suggests that *Pikaia* is an early chordate offshoot rather than one of our direct ancestors. The Chengjiang fauna is somewhat earlier in the Cambrian than the Burgess Shale. If there really were vertebrates alive then (about 525 MYA), at least one split between vertebrates and non-vertebrate chordates had already happened millions of years before the animals of the Burgess Shale swam in the Cambrian seas.

Because the Burgess Shale holds a special place in the history of the study of Cambrian animals, this account of the Cambrian fauna would be incomplete without a brief mention of its discovery. All I've said so far is that the rocks in which the wonderful fossils of animals like *Anomalocaris* were first found are in British Columbia. But who discovered their secrets?

As with many historical stories, the more you probe at it the more players you reveal, and the more complex the story becomes. But here I'll concentrate on the man who can reasonably be described as the main player. Canadian geologists were already aware of the existence of Burgess Shale fossils by the 1880s, and perhaps earlier. But the key figure was an American palaeontologist called Charles Doolittle Walcott, who discovered the incredibly fossil-rich nature of the Burgess Shale rocks in 1909, collected tens of thousands of specimens over the course of several visits extending for 15 years (the last visit was in 1924) and set about interpreting them. It was a major part of his life's work. One of the main collecting sites has been named after him: Walcott Quarry.

Walcott died in 1927, and although several other people studied some of his material in the following couple of decades, the fossils were only dragged out of obscurity in the 1960s and 1970s by the Italian palaeontologist Alberto Simonetta and (independently) by Harry Whittington, who set in motion a series of new collecting trips that resulted in the acquisition of thousands more specimens. From there the story continues via Briggs, Conway Morris and Gould, as noted earlier, along with many other palaeontologists.

Is the story of interpretation and reinterpretation of the Burgess Shale fossils over? Almost certainly not. But the pace of change seems to have decelerated. This is perhaps a sign that we are approaching a consensus view – which of course may or may not be correct; in science, agreement on a theory does not mean that it's right. This is all-too-plainly evidenced by previous agreement on many theories in both the biological and the physical sciences. In terms of the relationships among living animals, when I was an undergraduate in the early 1970s the segmented worms such as the earthworm and lugworm were generally agreed to be very closely related to the arthropods, partly because they were both groups of segmented invertebrates. Much research time and energy was put into the question of how a segmented worm became 'arthropodized'. We now know that this consensus on a sister-group relationship between the two was wrong (see Chapter 18), and that the theories of the arthropodization of worms were misconceived. I suspect that the fossils of the Burgess Shale, along with the related fossils from Chengjiang and Sirius Passet, may still hold some surprises.

5　How to make a species

Charles Darwin's *On the Origin of Species*, published in 1859, can rightly be regarded as the start of evolutionary biology. That's not to say that it was the first publication on evolution, but it was the first to convince most scientists – in some cases immediately and in others eventually – that (a) evolution had happened and (b) it occurred via a particular mechanism, namely *natural selection*, or 'survival of the fittest'.

In the previous chapter we saw that, whatever uncertainties remain about the origin of animals, by 500 million years ago the Cambrian oceans were teeming with animal life. They were doubtless teeming with bacterial and algal life too, though the flowering plants that dominate the plant kingdom today did not evolve until much later. In the Cambrian, all multicellular life-forms were aquatic – the land did not get colonized by plants and animals for probably another 100 million years.

Because natural selection is a *general* mechanism of evolutionary change, it must have been operating in those ancient marine ecosystems of the Cambrian much as it operated in their more recent terrestrial equivalents over the last six or seven million years to modify human and chimp lineages from their last common ancestor. And indeed it is still operating in the same way today, as we see in the evolution of biocide-resistant insects and bacteria.

The origin of a new species – or *speciation* as we now call it – requires two things: (a) divergent changes (at least some of them selectively driven) in two or more geographically separate populations belonging, initially, to the same species; and (b) reproductive isolation of the two forms so that they become – by definition – species in their own right. In such a scenario, we usually speak of a parent species and its two daughter species. However, this account is a simplified one, and we need to probe into a few 'ifs and buts' to arrive at a more complete picture.

First, the appearance of new species does not *have to* start with geographic separation of two or more of its constituent populations – but this is regarded as the norm, at least in animals. So the American biologist Richard Lewontin may have put the case too strongly when he said, in his 1974 book *The Genetic Basis of Evolutionary Change*: "If there is any element of the theory of speciation that is likely to be generally true, it is that geographic isolation and the severe restriction of genetic exchange between populations is the first, necessary step."

Second, 'selectively driven' does not capture the entirety of the reason for the divergence, though it's fine as far as it goes. The basic idea is that if two populations occupy separate environments where ecological conditions are different, natural selection will favour different variants in the two places, thus driving population divergence. But where does the variation that selection acts on come from in the first place? Genetic mutation is part – but only part – of the answer. And is the variation entirely 'random', as it has often been portrayed, or is it structured in some way (as Darwin thought)? If the latter, what effect does this have on the nature of the divergence? We will leave these questions open for now but will return to them in subsequent chapters.

Third, reproductive isolation may take place in one of two ways. The simpler of the two occurs when the populations concerned remain geographically separate for so many generations that they accumulate sufficient genetically based differences that, whenever they eventually encounter each other again, they are so distinct that interbreeding is impossible. This may be due, for example, to structural differences in their genitalia or incompatible courtship behaviour. But suppose that the period of time spent in different environments is shorter and inter-breeding on re-contact is not impossible. In this case, there is often a problem with hybrid offspring in terms of inviability or infertility – the mule as a horse–donkey hybrid exhibits the latter problem. A special sort of natural selection then comes into play – selection against those members of each 'incipient species' that tend to engage in mating with the other. This kind of selection is called the Wallace effect, in honour of the co-founder, with Darwin, of the theory of natural selection – Alfred Russel Wallace.

Fourth, I have oscillated, in the above account, between talking about 'two populations' and talking about 'two or more'. In reality, it can no doubt be two, three, four, etc., all the way up to 'many'. The number depends on the structure of the environment. If the water level falls in a large lake of varying depth, the result will be a patchwork of small lakes, perhaps paving the way for several daughter species to arise in parallel from a single parental one. Likewise, when volcanic activity in the sea creates new land, it often produces many islands rather than one: this is true, for example, of the Galapagos Islands off the coast of Ecuador, the setting for the much-studied speciation of the birds that have become known as Darwin's finches.

Fifth, the two or more daughter species that are the 'end-products' of speciation are also the starting points for the process to begin all over again. They may start life as new species that are endemic to small areas, but if they are successful and spread, then they, like their parental species, will come to consist of a patchwork of populations – and these in their turn will begin to undergo selectively driven divergence. Thus the processes of speciation, geographic spread and selective divergence form a cycle, not a line. This cycle has been replaying itself over and over ever since life began.

Given this almost-infinite replay, we can begin to think about the bigger picture that will emerge from it; and we can link this to Linnaeus's and Darwin's concept of 'groups within groups'. Multiple speciation events from the starting point of a single stem species will produce a group of related species in a way that can be depicted as an evolutionary tree. But further speciation events will produce an even bigger tree; and the extinctions of some of the species – an inevitability – will produce gaps in the tree. In some trees, all the branches will die out, leaving us at best with fossil evidence from which to try to reconstruct the tree of descent that is no longer apparent from any living animals.

The business of how to reconstruct evolutionary trees in general – not just those in which all the species are now extinct – will be dealt with in Chapter 7. For now, we will simply assume that there is a way to do it; and we will return to a sharper focus on the more restricted issue of 'how to make a species'.

Where better to start than with Darwin's *On the Origin of Species*? In his chapter on natural selection (chapter 4), Darwin begins with an explanation of the selective process. He then goes on to say: "This preservation of favourable variations and the rejection of injurious variations I call Natural Selection." However, careful as always not to overstate his case, Darwin continues: "Variations neither useful nor injurious would not be affected by natural selection, and would be left a fluctuating element, as perhaps we see in the species called polymorphic."

The term *polymorphic* refers to the existence of two or more distinctly different forms of a species in the same geographical area. Such different forms may occur at various levels, from the molecular to the organismic. Ironically, Darwin is probably right that some polymorphisms are selectively neutral, as we now call it, though selective neutrality tends to occur more readily at the molecular level; and molecular polymorphisms (e.g. in the structure of proteins) were unknown in his time. In contrast, it turns out that many higher-level polymorphisms affecting the whole animal, or a large part of it, *are* subject to selection after all.

To illustrate this point, and indeed to illustrate selection in general, I'm going to use snails: to be specific, two species of 'banded snails' called *Cepaea* (Figure 5.1). The two species are called the brown-lipped and the white-lipped banded snails – though always beware 'popular' names for animals because they are often region-specific and also can be misleading. Their scientific names are *Cepaea nemoralis* and *Cepaea hortensis*. Those of us who have done research on them affectionately call snails of these two species 'nems' and 'horts'; I'll use these abbreviated nicknames here.

Both of these species are polymorphic for the colour and banding of their shells. They can be adorned with black bands or not (Figure 5.1); the bands, when present, can also vary in number – up to a maximum of five. The paler, non-banded part of the shell can also vary in colour. Most if not all populations of both species are polymorphic in these respects, and the relative frequency of the different forms varies from place to place. Let's focus first on snails of the more-studied species: nems.

Careful study of the pattern of variation among populations living in different places has revealed natural selection at work. One example: it has been shown that the frequency of paler shells that take up less heat increases from the cold climates of Scandinavia to the hotter ones of Mediterranean regions. Stating that snails with paler shells are at a selective advantage over those with darker shells in hot climates is the same as saying that they are *fitter* in those climates. In evolutionary biology, the fitness of one variant in a population relative to another refers to it having a higher probability of surviving to adulthood and/or a greater likelihood of having offspring.

But why, you might ask, does the fittest type for each region not end up comprising the whole of the population there, as opposed to just

FIGURE 5.1 Two species of land snail: *Cepaea nemoralis* (top) and *Cepaea hortensis* (bottom). Snails of both species are polymorphic for shell colour and banding. The bands, when present (left) are more darkly pigmented than the rest of the shell; indeed, they are often almost black. On average, individual snails belonging to *C. nemoralis* are larger than those of *C. hortensis*; they also have a dark rather than a white lip.

having a high frequency? The reason seems to lie in another kind of selection that is less well known than the 'ordinary' natural selection that we now describe as *directional*. This other kind of selection is *balancing*. Here, natural selection causes not the elimination of a less fit form from a population, but rather the production of a state of balance between two or more forms. At first sight this seems puzzling: how could selection work in this way? Well, we have evidence for how it might do so both in this particular example – snail-shell pigmentation – and in others.

Land snails are eaten by thrushes and other birds. Thrushes have even devised a clever technique for smashing the shells to reveal the soft, edible flesh within. They drop the snails on stones – called anvils – and watch the shells break. They then peck at and consume the snail's de-shelled body.

It seems that birds form 'search images' for their prey. That is, they have a kind of mental image of a snail as a potential food item. This influences their mode of hunting in that they tend to overlook snails that do not correspond to their search image. The search image is moulded on the commonest form of snail – this makes good sense, because there's no point in flying around looking specifically for some-thing that is rare. So the birds often neglect – not completely, but comparatively – rarer forms. The rarer a particular form gets, the more it is neglected; thus its frequency increases because in this respect – low susceptibility to predation – it is fit. But as its frequency increases it loses this advantage for the very same reason that it acquired it when rare. The end-result of this kind of selection, which has been termed apostatic selection by the British population geneticist Bryan Clarke, is a state of balance between the different forms.

So this example shows how natural selection operates in natural populations, even in cases where Darwin suspected that it might not. But what does it tell us about the origin of species?

To make this link we have to dig deeper. Quite apart from the polymorphic variation in shell colour and banding on which selection acts in nems (and likewise in horts, though this species has been less studied), selection has also been acting on other characteristics of these snails. You might have noticed in Figure 5.1 that the nems are depicted as being slightly larger in shell size (and hence body size) than the horts. This is not accidental – it's true that on average nems are bigger. You may also have noticed that, as expected from the common names of the two species, nems have dark lips to their shells while horts have white lips. In addition, some internal characteristics have diverged between the two species (they're just harder to see). So although some of the selection acting on nems and horts has been of the balancing kind, some has been of the more familiar directional kind, and has been responsible for the eventual production of the two daughter species that we find today from a single, earlier, ancestral species.

We should pause at this point and ask: is it certain that what we call nems and horts are in fact separate species? These days, a species is defined in terms of reproduction: two animals are members of the same

FIGURE 5.2 The liger hybrid: the consequence of a mating between a lion and a tiger. There is an important difference between such hybrids produced in captivity on the one hand and those produced in nature on the other. Lack of hybridization in nature is the criterion that is ideally used to delineate animal species.

species if they can reproduce with each other in nature – usually with the proviso that they are of opposite sexes. Strangely this 'opposite sex' thing is inapplicable to *Cepaea* land snails, which are cross-fertilizing hermaphrodites. Note the 'in nature' caveat: you may already be aware of the kind of animal called a liger (a lion-tiger hybrid, Figure 5.2), but this creature was not the result of reproduction in nature – hardly surprising, given that the geographic ranges of lions and tigers don't overlap.

If you head out into some appropriate habitat, such as the limestone dales of Derbyshire, early on a damp May morning, you'll see mating pairs of *Cepaea* snails. If you look closely, noticing in particular the lip colours, you'll see that most of these pairs are composed of two snails of the same species, with one hermaphrodite playing the male role, the other the female one. So most mating pairs are nem + nem or hort + hort.

But occasionally you see a pair that is nem + hort. So a member of one species is attempting to copulate with a member of the other. The question is: do they succeed?

We know, from a meticulous, large-scale study of *Cepaea* undertaken by a German biologist more than a century ago, that the answer is 'yes'. Looking at fine detail of the structure of certain internal parts of the snails, as this biologist did, it seems that, in an area where both species coexist, hybrids constitute about 1% of the overall population. What we don't yet know is whether they are fertile or, as in the case of the mule, sterile. Regardless of this, what we see in the very limited mating of nems and horts is the achievement of the 'species barrier' but in a still-imperfect form, because the divergence of nems and horts was recent in evolutionary terms.

Now we have a reasonably clear picture of what is going on: despite balancing selection acting on some characters in both species, directional selection has been acting on others, with the result that nems and horts are now incapable – or nearly so – of interbreeding. However, there are still many open questions. For example, what was the geographic pattern of the speciation event? And is the Wallace effect now at work as a special form of natural selection, reinforcing the near-complete reproductive isolation of the two species so that it will become total? There is much left to study in the evolutionary biology of these fascinating snails.

As noted earlier, the processes of geographic spread of a new species, selectively driven divergence in its constituent populations living in different places, and reproductive isolation, form a cycle rather than a line; and several rounds of the cycle produce a tree of closely related species. Although nems and horts are the only two species of *Cepaea* found in the British Isles, two more species are found in mainland Europe. And *Cepaea* species form only a small part of the family Helicidae, which has hundreds of member-species worldwide.

The beauty of Darwinian natural selection – and the 'speciation cycle' of which it is a part – is that it applies to all species at all times throughout evolution. So, in broad terms, the type of scenario described above for one particular group of snails is played out time and again. It was doubtless responsible, long ago, for the diversification of the many species of *Anomalocaris* that have now been discovered in addition to the original one discussed in the previous chapter: these, like any group of related animals, can be arranged as a tree. And equally, it was responsible for the evolution of the tree of species of *Homo* that we'll look at in Chapter 24. Of course, with extinct species there are many things that can't be directly studied – like the process of mating discussed in relation to *Cepaea* snails. But there is no reason to believe that selection acted any differently in extinct species than it does in extant ones.

There is, however, a potential problem when we think even more broadly across the animal kingdom. That Darwinian selection, coupled with reproductive isolation, can cause new species to originate within a group of reasonably similar animals is no longer in doubt among scientists – except to 'creation scientists', who don't count because they're not scientists at all. So we have no difficulty in understanding the proliferation of slightly different species of giant Cambrian predators, of diminutive European land snails, or of African proto-humans. But the really interesting question is whether, by extending the cycle of speciation over the aeons, we can link all these small trees up into one big tree – the tree of all animals.

Here we have come face-to-face with the issue of whether, as some have put it, evolution is 'scale-independent'; in other words, whether the processes described above for snails can, when 'writ large', explain even the deepest splits in the animal kingdom. There is not yet a consensus about this issue among evolutionary biologists. Some see a seamless continuum from small, shallow divergences, such as between the different species of *Cepaea*, to large, deep divergences, such as the one between their phylum (the molluscs) and ours (the chordates). Others see a need for *something else* in the explanation of the deep splits. We'll revisit this issue later, for example in Chapter 23 in relation to the origins of what are known as evolutionary novelties – such as the shell of turtles.

One of the holders of the second of the two above views – the 'something else' view, if you like – was the great Scottish biologist D'Arcy Thompson, who wrote his magnum opus, *On Growth and Form*, in 1917 – almost a century ago. In it, he used a geometric approach to show how closely related species often have forms that can be connected by what he called a transformation (Figure 5.3). But precisely because he knew that his method could *not* be applied to very distantly related

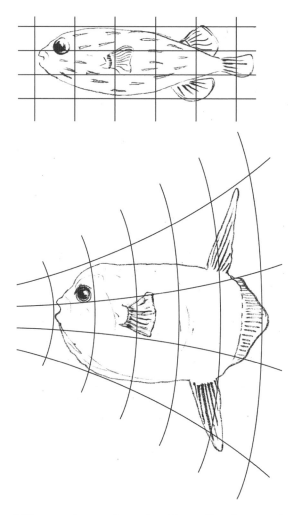

FIGURE 5.3 A Thompsonian transformation. The outline of a porcupine fish (top) can be transformed into that of a sunfish (bottom) by a distortion of the grid on which the first outline was drawn. D'Arcy Thompson argued that evolution might act by transforming whole developmental trajectories in this way, rather than by acting separately on lots of different characters. However, the transformations that Thompson used as examples, including this one, were for the purposes of general illustration and should not be taken to imply ancestor–descendant relationships between the species concerned.

species, such as those from different phyla, he stated that his geometric analogies "help to show that discontinuous variations are a natural thing, that 'mutations' – or sudden changes, greater or less – are bound to take place, and new 'types' to have arisen now and then."

Interestingly, this is at odds with Darwin's views on the subject. Darwin frequently stated that *"natura non facit saltum"* (nature does not make jumps). Thus he took the first of the two views described above and saw a continuum from small, shallow divergences to large, deep ones. Perhaps the resolution of this apparent difference of opinion can be found in the modern approach of 'evo-devo' (evolutionary developmental biology), which we will examine in subsequent chapters.

6   Jellyfish and their kin

In the previous chapter, we saw that many rounds of the speciation cycle from the starting point of a single stem species produce many species whose pattern of relationship can be illustrated in the form of an evolutionary tree. As time goes on, the number of extant species that constitute the growing tips of the tree's branches increases, unless the number of species-deaths – through extinction – becomes large enough to balance the number of species-births, in which case the number of species stabilizes. If the number of deaths exceeds the number of births, then of course the size of the overall group of species declines.

When we look at a group of related species alive today, it's natural to wonder what their stem species, or last common ancestor (LCA), was like. This often motivates searches for fossils representing possible candidates. However, while candidates can indeed be found, ascertaining whether any one of them actually was the stem of the group concerned is difficult. This is true even of small groups of species with a relatively recent origin – such as the great apes. It's even more of a problem when dealing with large groups of species with an ancient origin.

In this chapter, I want to discuss a large group of animals that diverged from the rest of the animal kingdom shortly after its inception, in terms of geological time. This group includes an array of at-first-sight disparate creatures such as jellyfish, sea anemones and corals. The group is at the phylum level, which, as you'll recall from Chapter 4, is the highest conventionally recognized level of group beneath kingdom. This chapter is the first of several to focus on a particular phylum with a view to acquainting readers with (a) the uniting features of the group of animals concerned and (b) the diversity that often obscures such features and yet is of considerable interest in its own right.

The phylum in which jellyfish and their allies are included is called the Cnidaria (with a silent C so that it is pronounced as if it starts with an N – nigh-dairy-ah). This name is related to a common feature of all animals within the group – the presence of stinging cells called cnidocytes (again with a silent C). Indeed, the presence of these unusual cells (Figure 6.1) is one of the things that persuaded zoologists to unite

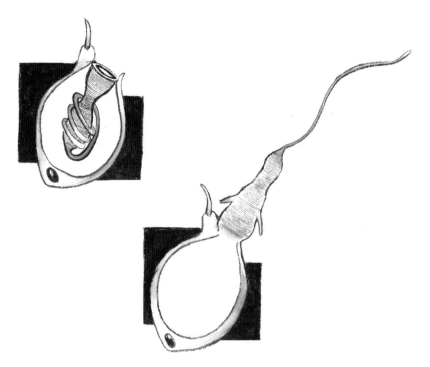

FIGURE 6.1 The cnidarian stinging cell or cnidocyte, shown in two forms: ready to be fired at a prey item (left) and having been fired (right). The upward projection has a sensory function: contact between it and a potential prey item is the trigger for firing.

such different-looking creatures in a single group long ago. Now we have confirmation of the correctness of their decision in the form of DNA sequence data. So the Cnidaria is indeed a 'natural' group. And it's a rather large one: it includes more than 10,000 extant species.

The great differences in size, colour and general appearance among these 10,000 or so species conceal a common growth form and a common life-cycle. Cnidarians do not, like us, have a digestive system with two openings – mouth and anus – to the outside world. Rather, they just have one, which thus has to function for both inlet of food and egress of wastes – an idea that we find alien, and indeed unsavoury. However, regardless of our subjective views on this arrangement, it clearly works well enough – otherwise the group would not have expanded to its current size.

The general cnidarian growth form can be thought of as having just three main features: the digestive cavity (with its single opening), the

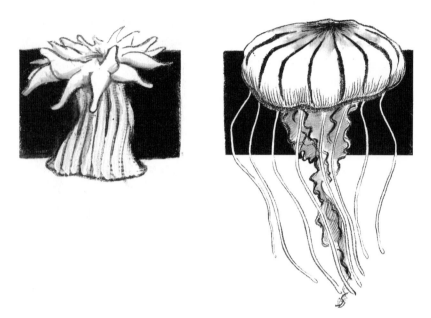

FIGURE 6.2 Two cnidarian growth forms: the polyp form of a sea anemone (left) and the medusa form of a jellyfish (right). Anemones are typically attached to a hard substratum, such as a rock; jellyfish, of course, are free-swimming.

quasi-cylindrical or dome-like envelope of tissue that surrounds it (typically consisting of two cell-layers), and multiple tentacles protruding from around the digestive opening, which bear the stinging cells. These cells are used to attack prey items, the nature of which varies depending on the size of the cnidarian.

Although I have described the growth form as a general one – in the sense of applying to all species in the group – there are two versions of it. The difference between these lies in which way up the whole structure is – mouth at the top or mouth at the bottom. Associated with this difference is another: the mouth-at-the-top form is usually attached to the substratum, while the other form is usually free-swimming. And, as an aside, substratum and swimming usually refer to the sea, since the vast majority of cnidarians are marine, though there are a few that live in fresh water; none, however, live on land.

The two orientations are referred to as polyps and medusae (singular, medusa; Figure 6.2). Polyps share their name with some tumours in humans that have a roughly cylindrical growth form. The name medusa comes from Greek mythology. One of the Gorgons – a type of monster – was a woman so ugly that those who looked upon her turned to stone. This creature, Medusa, was crowned with a writhing mass of serpents in place of hair. She was beheaded by Perseus, who approached her warily, looking at her only indirectly, via her reflection in his polished mirror-like shield, rather than directly, so that he would avoid the turning-to-stone fate. Medusa's severed head, with snakes dangling from around its circumference, is a bit like the mouth-at-the-bottom orientation of the cnidarian growth form.

Having separated the two versions of this growth form, we'll now bring them back together again. This requires looking at them as four-dimensional rather than three-dimensional things. In other words we are back to the perspective, espoused in Chapter 1, of thinking about animals as *being*, rather than just having, life-cycles. In the Cnidaria, this perspective is especially important.

Most of us have seen sea anemones, which are examples of the polyp growth form. Also, most of us have seen jellyfish, examples of the medusa form. Indeed, to anyone who lives near the sea, both are familiar sights. But we don't immediately spot a connection between them. Anemones are often vividly coloured and seen in rock pools; many of the jellyfish found in temperate regions are very pale, almost anaemic looking, and are seen in open water, perhaps from a vantage point at the end of a pier. They both have tentacles, but even these don't look very similar – those of the jellyfish are much longer and somehow manage to inculcate into human observers a sense of danger not so readily picked up from observing their anemone cousins.

Now here's a thing that not a lot of people – well, not a lot of non-biologists really – know. Many cnidarian species have a life-cycle that alternates between polyp and medusa stages. The polyp produces – by asexual means – little buds or slices within itself which detach and float away from their parent as medusae. Eventually, the medusae produce eggs and sperm, which are usually released into the surrounding seawater, where fertilization takes place. Each fertilized egg divides many times, eventually resulting in the formation of a small larva called a planula, which will settle somewhere on the substratum and turn into a polyp, thus completing the cycle, as shown in Figure 6.3.

As ever, though, nature is complex and varied. Some cnidarians have only a polyp stage while others have only a medusa stage. And of those that have both stages, some have one or the other very much in

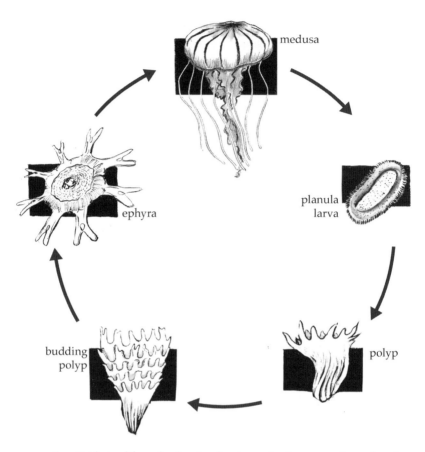

FIGURE 6.3 Cnidarian life-cycle, showing the alternation between polyp and medusa stages that occurs in many of these animals. Sexual reproduction of the medusa stage produces a small larva, the planula, which settles and develops into a polyp. Asexual reproduction by the polyp takes the form of a series of slices, each separating off as a stage called the ephyra. Each of these then develops into a medusa. (The plurals of planula, ephyra and medusa all have a terminal *e*: planulae, ephyrae and medusae.)

the ascendancy in that it is the larger of the two and/or extends in time for a much greater portion of the life-cycle. Then again, there are some cnidarians where the life-cycle is quite balanced between polyp and medusa stages.

Of course, this variation among living species raises the question of whether, going back in time to the cnidarian stem, the original form was a polyp, a medusa, or an alternation of the two. At present, there are arguments and counter-arguments, but no consensus on the answer to this fascinating evolutionary question.

Another evolutionary question, this time one on which there is a consensus, is how the cnidarians are related to other branches of the animal kingdom. We already saw that sponges are primitive animals – it's very hard to break the habit of using 'primitive' and 'advanced' although, as noted in Chapter 1, we must avoid the associated picture of the animal kingdom as a vertical series of creatures and evolution as an escalator taking each of them upwards at different rates. What we really mean when we say that sponges are primitive can be broken down into two things: a low level of body complexity and an early evolutionary divergence point.

We'll look at complexity later – Chapter 11 is devoted to it. For now, suffice it to say that while 'advancedness' is hard (impossible?) to define or measure, complexity is not. One measure of the complexity of an animal is the number of different cell-types it has. Sponges have very few – hence they are said to have a low level of complexity. Regarding earliness of divergence, sponges seem to have split from the lineage leading to the rest of the animal kingdom earlier than did any other major group.

It might be sensible to dissect the word *major*. In this book, I am trying to keep things simple and to keep terminology – including that pertaining to the naming of animal groups – to a minimum. One way to help achieve these aims is to avoid mention of minor animal phyla in the sense of those that include very few species. As we saw above, there are over 10,000 species of Cnidaria, so this is a major phylum. Sponges, with about 8000 species, can also be considered major. But there are other primitive (i.e. low-complexity, basally branching) animal phyla that contain only about 100 species, and indeed at least one such phylum that contains just a single species. If you want to learn about all of these, a standard textbook on invertebrate zoology – such as the 2003 book *Invertebrates* by the Brusca brothers – is the best source.

Anyhow, continuing with the simplified version, here's what we think happened a long time ago. After the origin of animals, sponges split from 'the rest'; then cnidarians did likewise – see Figure 6.4, which also shows alternative possible branching patterns that are thought not to have happened.

Part of the reasoning behind preferring the branching pattern shown at the left of Figure 6.4 concerns the possession, or otherwise, of a nervous system. Cnidarians and other animals have such systems but sponges lack one. Thus the nervous system is said to be a shared derived

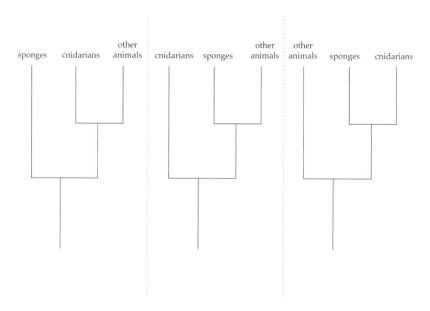

FIGURE 6.4 Three possible evolutionary branching patterns in the early animal kingdom. The one thought to be correct is on the left, with sponges branching off first, then cnidarians. There is a fourth possibility, namely that the sponges do not comprise a natural, or monophyletic, group; however, at the time of writing, the balance of probability does not favour this.

character between cnidarians and other animals (including us). This fact suggests that the sponges split from the lineage leading to nerve-bearing animals before that lineage split further. So the nervous system is *informative* about the pattern of branching. In contrast, multicellularity is not an informative character, because all three groups shown in the figure exhibit multicellularity, a feature that they presumably all inherited from the stem animal. (As far as we are aware, no lineage of animals has ever reverted to unicellularity.)

The person responsible for articulating this use of shared derived characters was the German palaeontologist Willi Hennig. His book *Phylogenetic Systematics* – a 1966 translation of his book published in German in 1950 – laid down the groundwork for a more objective approach to determining the correct arrangement of groups within groups than had prevailed before. Pre-Hennig, the approach taken was too dependent on the intuitions of specialists in the groups concerned.

Having an objective approach to inferring the branching pattern of evolutionary trees is definitely an advance; but, as ever, nature conspires

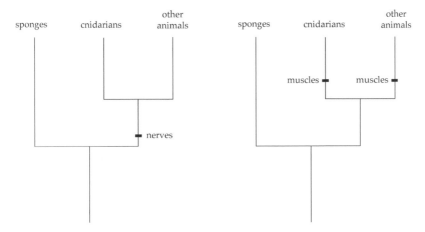

FIGURE 6.5 Different mappings of nerves and muscles to the left-hand tree of Figure 6.4. As can be seen, nerves are thought to have originated just once, but muscles twice. If this is true, then our muscles and those of a jellyfish are the result of convergent evolution.

to make our job difficult. To see this, we only need to move from nerves to muscles. Sponges lack these, while both cnidarians and 'other animals' have them. But subtle differences in the structure and embryonic origin of our muscles and those of cnidarians suggest that these were independently – or convergently – evolved. So when a jellyfish swims, it is using a kind of muscle that probably first arose in the stem cnidarian. Figure 6.5 shows the difference between nerves and muscles in terms of how their origins are thought to map to the early branching of the animal evolutionary tree.

There's an important group of cnidarians that I haven't mentioned yet: corals. Within the Cnidaria, these are quite closely related to sea anemones. This seems strange at first. Corals are usually thought of in the form of coral reefs: how can something that's large and hard be closely related to something that's small and soft? It's a bit like finding out that earthworms are closely related to lobsters (they're not). The answer lies in corals having two features that the typical sea anemone does not – colonial growth and secretion of a kind of exoskeleton – but also having many features in common with anemones.

Coral reefs consist of many millions of coral polyps, which normally reproduce asexually, plus the hard, stony substance that they make, which is a crystal called aragonite – one of the three crystal forms of calcium carbonate. Each polyp secretes a plate of aragonite, and these join up to form the large stable structure that we call a reef.

Reefs are distributed mostly in shallow tropical and subtropical seas. They are famous for being a sort of marine equivalent of the terrestrial environment's tropical rainforests in the sense that they are home to extremely high levels of biodiversity. Coral reefs harbour far more species per unit size than do any other marine habitats, such as a stretch of rocky shore or a given area of sandy substratum. Indeed, it is frequently the case, in TV natural history programmes about reefs, that the various non-cnidarian species that live in a reef – including forms as diverse as fish, sea slugs and sponges – get more attention than the cnidarians that built the reef and are at its core.

The reason that most reefs occur in shallow water is that the majority of coral species get part of their food supply from symbiotic algae. These need light in order to photosynthesise, and, below a certain depth in the ocean, the light is too dim. But, having said that, corals do have the typical cnidarian stinging cells noted earlier. They can capture prey using these, and also by using mucus traps.

I've managed to get this far in my story of this fascinating group of animals – the Cnidaria in general rather than just corals in particular – without emphasizing a central theme of their growth form: its type of symmetry. This has been implicit in the illustrations, in particular Figure 6.2, but it will now become explicit.

One way of approaching this topic of body symmetry is through the nervous system. We already saw that cnidarians are more complex animals than sponges, in that they have such a system while sponges do not. However, cnidarians have a less complex nervous system than most other animals do, in that they lack a brain. Instead of having a central and a peripheral nervous system, as we do, they only have a diffuse nerve net. As an aside, here, an interesting puzzle is how the group of cnidarians called box jellies (which are even more poisonous than 'ordinary jellyfish') can see. These strange creatures have remarkably sophisticated eyes and yet no brain to interpret whatever images they are capable of forming.

Having no brain is part of having no head. Although it's tempting to see the mouth-end of a polyp as the head, you have to remember that the mouth is also the anus – or alternatively that it's neither. So it can't be used to determine the head end any more satisfactorily than the nervous

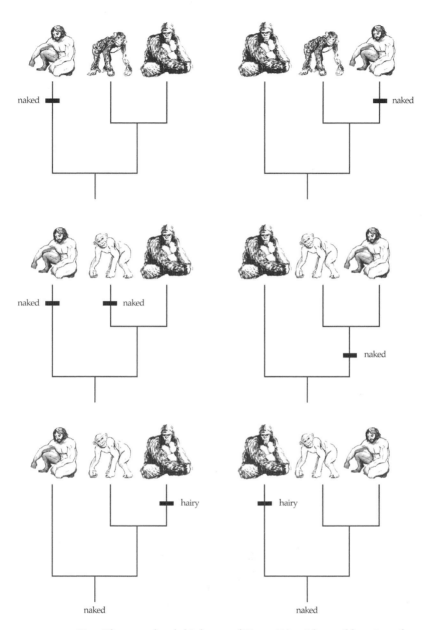

FIGURE 7.2 Top: The second and third trees of Figure 7.1, with possible points of transition from hairy to naked shown. Centre: The same two trees, but for the hypothetical situation in which chimps are naked rather than hairy. Bottom: Another hypothetical situation in which chimps are again naked but the last common ancestor of humans, chimps and gorillas was naked too.

than Occam took a similar approach, but somehow Occam ended up getting the credit for a general principle that we often use in our daily lives: simpler explanations are generally preferable to complicated ones.

Now, back from the general to the particular – the choice between two competing evolutionary trees. To see a situation in which there *is* a difference between two alternative trees of human–chimp–gorilla relationships in terms of how parsimonious they are, we will move to an imaginary world where chimps are naked. In such a world, the equivalents of the two real-world trees shown at the top of Figure 7.2 are shown immediately beneath them.

Notice that we now have a different situation: one tree – the one with humans and chimps as sister-groups – can produce the pattern of character states that is found in our fantasy world with a *single* evolutionary shift in character state; the other tree – with chimps and gorillas as sister-groups – needs *two* such shifts, and is thus less parsimonious.

There is, however, a problem with this approach that has so far remained hidden. We have assumed that the extinct creature which the stem of all these trees represents was hairy. If in fact this ancestral creature was naked, albeit that's less likely, then the two alternative trees in the imaginary naked-chimp world are equally parsimonious (see the bottom panels of Figure 7.2).

The reason why the stem creature is assumed to have been hairy is also due to the use of Occam's razor. Given the generally hairy nature of the gibbons and the monkeys, it seems most reasonable that the lineage leading off to the great apes was also hairy. Here, we are using a form of reasoning that includes information not just on the three species that are the focus of our attention but some outside it too – *outgroups*, as Willi Hennig called them. And I should now re-emphasize a point made in the previous chapter: we have been using another concept to which Hennig gave a name: *shared derived characters* as opposed to shared primitive ones. In the world of the naked chimp and the hairy ape ancestor, humans and chimps share their nakedness not because it was the original character state of the primates in general but because it arose in the last human/chimp common ancestor.

There's quite a lot in the last few paragraphs, but at least I've managed to keep the terminology to a minimum. For those readers who would like to find out a bit more about Hennig and his work, try Googling the following terms: symplesiomorphy, synapomorphy, autapomorphy. For those who would prefer not to, don't worry – none of these terms is necessary for readers of this book. Hennig did the world of evolutionary biology a big favour by articulating an objective approach to working out the correct evolutionary tree for any given group of animals. But – at least in my view – he invented too many complicated-sounding terms.

Now let's return to the simpler of the two problems that we started with. Since that was a few pages back, here's a reminder: hairiness is just one character. If we look at other characters, they may suggest different trees. Possible choices might include erect posture with bipedal walking and the possession of large brains – say those whose volume is more than 1000 cm^3. Both of these give us two equally parsimonious trees for human–chimp–gorilla relatedness, just as hairiness does (the middle and right-hand trees in Figure 7.1). The reason is the same – quadrupedal locomotion and small brains are both primitive character states, not the shared derived ones that we need to use to build trees.

So what characters do distinguish between alternative trees and put humans and chimps as sister-groups using a valid rationale? Most of them are molecular characters. Some of the earliest molecular work was done on the structure of the haemoglobin molecule (and other related proteins of the globin group) that carries oxygen in the blood of humans, chimps and gorillas, and of course in the blood of other vertebrates too.

This early work showed that uniquely shared derived globin sequences were more common between humans and chimps than between chimps and gorillas, thus arguing that humans and chimps are sister-groups. Such early work on particular proteins, and the genes that make them, has now been complemented by whole-genome studies; and the results of these are the same. So, the human–chimp sister-group relationship can finally be regarded as having been established beyond reasonable doubt.

I do not want to imply, here, that molecular characters are always better than morphological ones (i.e. those relating to an animal's structure) for finding out the correct tree – either for the great apes or for any other group of animals. Sometimes they are, sometimes not. But the good thing is that, now we have information on both kinds of character at our disposal, we can use both and see if they tell the same story. If they do, then the tree that they suggest is almost certainly the right one. If they don't, then we have to look carefully at the way the two types of evidence conflict, and try to resolve the problem – perhaps by determining the states of more characters of both types.

Given that it has taken decades of study to determine the correct evolutionary tree for the great apes, with humans and chimps being finally recognized as sister-groups, imagine how difficult it is for students of bigger groups of animals, with many more constituent species. Even in the great apes, the situation is more complex than I have described it. For example, including just extant forms, 'chimp' covers two species – the common chimp and its close cousin, the pygmy chimp or bonobo. Also, 'gorilla' is now thought to include two species – the eastern and western ones (each of which in turn is thought to be divided into subspecies, though this latter term, unlike species, has no clear definition).

To *really* see the complexity that can confront a zoologist interested in the evolutionary relatedness of species within some bigger group, we should cast our net more widely. We could do this by simply extending outwards from our ape-based starting point. Thus, we could take all the primates and try to establish the correct tree for these – there are about 500 extant species of them. Or we could be more adventurous and take an entirely different group – say the turtles (including the terrestrial tortoises), which number about 200–300 species. Then again, we could venture out into the vast world of the invertebrates. Here, we might take the centipedes, as a 'manageable' group compared with, say, the insects – though even focusing only on centipedes means studying a group of animals with between 3000 and 4000 species, and thus a huge number (many trillions) of possible trees.

Often, those zoologists dealing with large groups such as these have to be content with resolving the evolutionary relationships between the major subgroups involved, for example those that would traditionally be called families, rather than aiming to end up with a tree that has a branch leading to every single species. Even that is a major undertaking. And I have deliberately chosen three groups – primates, turtles and centipedes – where both the group itself and its major subgroups are known to be monophyletic. We came across this term before, but just as a reminder it refers to groups that include all the descendant species that have radiated out from a single stem, or ancestor, and no others.

Zoologists working on animals where the overall group or some of its major subgroups may not be monophyletic have an even more difficult

FIGURE 7.3 A pill millipede, so called because it can roll up into a ball (right). The specimen shown is about 1–2 cm long and is from the northern hemisphere. In the southern hemisphere there is a group of giant pill millipedes, some of which are about 10 cm in length. The ability to roll up into a ball appears to have arisen independently in the two groups. Thus an overall group simply called 'pill millipedes' would not be monophyletic.

task. For example, someone specializing not on centipedes but on their close relatives the millipedes, of which there are more than 10,000 species, is faced with a situation where there is still not agreement on what the major subgroups are. Some currently and/or recently recognized groups of millipedes are probably not monophyletic (Figure 7.3). Eventually, the inclusion of molecular data (there is currently some, but very little) alongside morphological data may help to improve the situation, but this will not happen overnight. Indeed, it might not happen at all: although whole genomes can be obtained more easily and cheaply than in the past, each one still involves a considerable investment of time and money. If a particular group of animals is relatively obscure and contains few or no species of medical or economic importance, then it will be at the bottom of the list for genome projects. So we must face the very real possibility that for a long time many branches of the animal tree of life will remain shrouded in uncertainty.

Let's take a quick look at the way biologists refer to determining the correct tree for any group of creatures. It's often called tree reconstruction. This sounds odd at first to those who are new to the field, but there's a simple explanation. The idea behind this term is that when the group of animals concerned evolved – when they radiated out from their last common ancestor – they were effectively constructing their own tree. Our task, then, as students of the group concerned, is to find out how the tree was constructed by the evolving animals. So we are attempting to reconstruct it. I have some reservations about this usage, because it sounds as if we are somehow doing an experiment in re-running evolution, which is not, of course, intended. That's why I prefer to use phrases like 'trying to determine the correct tree'. But terminology is a means to an end. As long as we know what we're trying to do, it often doesn't matter exactly which phraseology we use to describe it, providing we use it consistently.

Finally, what do we do about the 'large number of species' problem? Determining the correct tree for three species and a small number of characters can be done by hand – making a table of species with their various character states, drawing a small number of possible trees, and using the former to discriminate between the latter. But when we are dealing with hundreds of species and often also hundreds of characters, the number of possible trees becomes so large that computers become not just useful but essential.

So, the science of naming animals based on their evolutionary relatedness has come a long way: from its intuitive starting point, where experts took a subjective approach using morphological characters only, to its current status as a hard science with a logical framework, a new (molecular) data-set to add to the old (morphological) one, and a vast array of dedicated computer packages. Using all of these, much progress has been made in understanding patterns of relatedness among animals over the last few decades. Our current picture of animal relatedness, which is the result of such progress, is summarized herein at a level appropriate for a general readership. Anyone wanting an advanced, in-depth account of animal relationships should consult *Animal Evolution* by Danish zoologist Claus Nielsen, the third edition of which appeared in 2012.

8   The enigmatic urbilaterian

Evolutionary trees, such as those we looked at in the last chapter, are very abstract things. They take a complex process involving many animals, molecules, morphological characters, matings, migrations, speciation events and geography, and render this multi-level, four-dimensional complexity into a simple-looking picture on a single sheet of paper. For some purposes that's great. After all, when we are trying to understand a complex process, any device that captures the essence of the process in a simple diagrammatic way is incredibly helpful. But it can also be misleading; and whether it does indeed capture 'the essence of the process' can be questioned. I think this is especially true in relation to the extinct animal that has come to be called the urbilaterian (and nicknamed, somewhat tongue-in-cheek, 'Urbi').

What was this strangely named beast? Answers to this question can come in a variety of forms. In terms of the name, it uses the German prefix *ur-*, meaning first or original, and couples that with *bilaterian*, which is short for 'bilaterally symmetrical animal'. As we saw in previous chapters, the most basally branching animals either had little or no symmetry (sponges) or had radial symmetry (jellyfish and their kin). But most other animals that we see around us are bilaterally symmetrical, albeit with various degrees of imperfection. Therefore, at some point in the distant past, bilateral symmetry must have originated, and the first animal displaying it was the urbilaterian.

But that's just an explanation of its name. What did it really look like? When did it first appear, and where? Which extant animals with bilateral symmetry display a form that is most similar to their enigmatic last common ancestor? We can't satisfactorily answer any of these questions yet; but this fact has not prevented the publication of many papers dealing with the supposed nature of the urbilaterian. Its pivotal position in the animal evolutionary tree has resulted in the publication of many speculative accounts, despite the fact that in science speculation usually attracts disapproval.

Ideas about the nature of the urbilaterian differ especially in how complex it may have been. The debate is sometimes presented in a binary way as simple versus complicated forms. Although it's probably better to think of there being a spectrum of possibilities along a complexity axis, discussion of the simple and complicated views is still useful, because these can then be taken as identifying the opposite ends of that spectrum.

In the simple urbilaterian view, this creature is seen as being a bit like some of today's flatworms. But care is needed here because there are a lot of different kinds of extant flatworms, and most people haven't seen all of them – indeed many people haven't seen any of them. There are both free-living and parasitic forms; also, flatworms can be found in marine, freshwater and terrestrial environments. There are forms that are so flat and wide that they're not really worms – some of these are called flukes, e.g. the parasitic liver fluke. Then again, there are forms that are so worm-like that they're not really flat. Three examples of flatworm diversity (out of an estimated total of around 20,000 species) are shown in Figure 8.1.

To make matters worse, the flatworms probably do not all belong in the same group. They used to be treated as if they did, and the group concerned was given a name that is essentially a Latinized version of flatworms (Platyhelminthes). But it now seems that at least one sub-group of flatworms doesn't belong with the others: the acoel flatworms (Figure 8.2). One version of the simple urbilaterian hypothesis is that the first bilaterian animal was much like today's acoel flatworms.

Let's go back to trees for a moment. The tree shown in Figure 8.3 is like one of the trees shown in Chapter 6 (Figure 6.4, left-hand tree) but with 'other animals' replaced by 'bilaterians', and an addition to the tree indicating the radiation of different bilaterian animals.

FIGURE 8.1 Three types of flatworm, illustrating the variation in general body form that is found in this group. Some are almost round in cross-section (left) while others are wider and flatter (centre and right). Overall, there are about 20,000 species of flatworms.

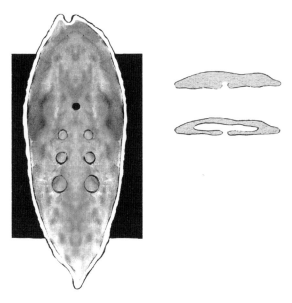

FIGURE 8.2 An acoel flatworm viewed from below (left), with the mouth darkly shaded. These rather featureless small animals are seen, in cross-section, to be solid (upper right), in contrast to members of the phylum Platyhelminthes, which typically have a gut cavity (lower right). In the acoels, food that is taken into the mouth is digested by a mass of cells specialized for that purpose.

The 'flatworm-like urbilaterian hypothesis' would be helped if we could be sure that today's acoel flatworms are the sister-group to all the other bilaterian animals. However, it does not *require* this to be the case, which is just as well because, at the time of writing, some zoologists think that they are such a sister-group while others think not. The issue remains to be resolved.

You might be wondering why the position of the acoel flatworms in the animal evolutionary tree is not necessarily a deciding factor. The reason is that convergent evolution is a very common thing – so it's entirely possible that the urbilaterian might have closely resembled today's acoel flatworms without having been closely related to them.

Many of the zoologists who favour a simple urbilaterian do so not mainly because of considerations about where the acoel flatworms fit on the animal evolutionary tree, but rather because of Occam's razor. You may recall this concept from Chapter 7, in which it was introduced: it is basically the idea of choosing the simplest, or most parsimonious, of competing hypotheses. And it's clear, indeed perhaps it's even

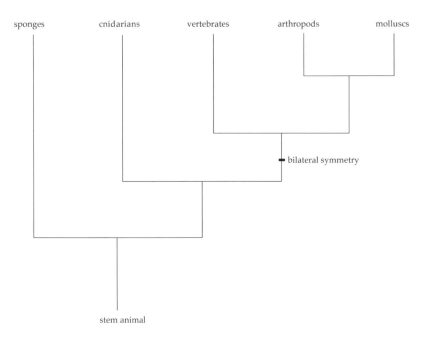

FIGURE 8.3 Evolutionary tree showing the pivotal position of the urbilaterian, the
first animal with bilateral symmetry, in the overall scheme of animal relationships.
Only three main lines of descent out of many leading from the urbilaterian are
shown – those leading to the vertebrates, arthropods and molluscs.

tautological, that the simple-urbilaterian hypothesis is simpler than the
complex-urbilaterian one. However, we can't get much further without
looking at the latter hypothesis in a bit more detail.

Supporters of this hypothesis envisage an urbilaterian that had many
features which the acoel flatworms do not, including a through-gut, a
segmented body, and many complex organs. If we concentrate on segmentation, we will be able to see the rationale behind the hypothesis.

In today's world, there are three main groups of segmented animals.
Members of our own group, the vertebrates, are segmented, but more
obviously so from the inside than the outside. This is because we have
internal skeletons, and segments are most obvious there in terms of our
spines being made of the segments that we call vertebrae. The muscula-
ture associated with vertebrae and ribs is also segmented, but you need
some anatomical training to appreciate this fully. From the outside we
don't look segmented because there is no trace of segmentation in our
outermost covering – the skin.

The arthropods are more obviously segmented because they have exoskeletons. So we can immediately see from purely external observation that arthropods are divided into segments along their head-to-tail axis. A close look will reveal different segmental patterns, such as the relatively fixed pattern of three thoracic and (approximately) 10 abdominal segments of the insects, in contrast to the relatively variable pattern of the centipedes. In a centipede, there is no distinction between thorax and abdomen, and the total number of trunk segments can be as small as 15 or as large as 191.

The third main group of segmented animals includes many species of worms. But these are not close relatives of the flatworms that we've discussed already in this chapter. Rather, these are the annelid worms – a group that includes the familiar earthworms, the leeches, and many species of marine worms, including those that leave worm-casts on sandy beaches and those that live in calcareous tubes affixed to rocks on the shore. Although these worms do not, like arthropods, have an exoskeleton, they are nevertheless obviously segmented from the outside, as anyone who has looked closely at an earthworm will agree.

Now the question arises of whether segmentation has originated once, twice, or three times in the animal kingdom. Back in the 1960s and 1970s, the answer to this question seemed obvious to most zoologists: twice. The annelids and arthropods were thought to be sister-groups, so it seemed reasonable to suppose that segmentation had been evolutionarily invented in their common stem lineage, which later split in two. Vertebrates, on the other hand, have never been thought to be very closely related to either the annelids or the arthropods, and their internal segmentation seems radically different to the kind of segmentation found in the other two groups. So the idea that it was a separate evolutionary invention seemed unassailable. This invention was thought to have occurred in the vertebrate (or perhaps the chordate) stem group. For a glimpse at the certainty with which scientists of those times viewed the double-origin theory of segmentation, see the 1964 book *Dynamics in Metazoan Evolution*, by English zoologist R. B. Clark.

Fast forward some forty years and certainty has been replaced by uncertainty. Most zoologists now reject the idea that segmentation has arisen twice in the animal kingdom. The reason for this is that the addition of molecular data to the evidence available for determining the pattern of evolutionary relationships among animals has revealed that annelids and arthropods are not sister-groups at all. Indeed, they are quite distantly placed on the animal evolutionary tree – almost as distant from each other as either is from vertebrates. But does this mean that segmentation has originated once or three times?

Another finding from the last forty years has added to, albeit not resolved, the debate. It turns out that although not all of the genes involved in making segments are the same in vertebrates, annelids and arthropods, some of them are. It can be argued that this would not be expected if segmentation had arisen independently in each group. Such an argument points to a single origin. But to complicate matters it is now clear that independent origins of a structure or series of structures – in this case segments – can be based on independently deploying the same genes, as long as these are present in the genome. This fact might be taken to imply three separate origins, especially when taken in conjunction with another fact – that most of the animals which are interspersed between the three segmented ones on the animal evolutionary tree are unsegmented. So, if there was just a single evolutionary origin of segmentation, there must have been a lot of evolutionary losses of it, which seems implausible – again on the basis of using Occam's razor.

To summarize the current state of play: There may have been a single origin of segmentation coinciding with the origin of bilateral symmetry; this is another way of saying that the urbilaterian was segmented. This view is expounded by the UK-based Spanish biologist Juan-Pablo Couso in a 2009 review article. Alternatively, there may have been three independent origins, in the stem lineages of annelids, arthropods and chordates, as argued by Israeli biologist Ariel Chipman in another review article published in 2010; in this case the urbilaterian was unsegmented. There is currently what I would call a stalemate in the contest between supporters of these alternative views.

All of the discussion so far has been based on relating the urbilaterian to extant animals. Perhaps we can shed some light on the issue by looking at fossils. Fossils may also help in terms of the question of when (and where) the urbilaterian lived.

As we saw in Chapter 4, there were many kinds of animals in the Cambrian period; and most of them, for example the predatory arthropod *Anomalocaris*, were bilaterally symmetrical. This suggests that the urbilaterian lived either at the start of the Cambrian or, more likely, earlier. But 'earlier' takes us into the Ediacaran period, which extended from 635 to 542 MYA. The problem with fossil creatures from that period, as we discussed in Chapter 2, is that there is not agreement about what they were, and, in particular, about whether they were animals or alternatively representatives of a failed experiment in multi-cellularity by a now-extinct kingdom.

We discussed previously the difficulties involved in interpreting several of the Ediacaran creatures, and there's no point in revisiting those arguments here. Anyhow, the nature of these creatures need not necessarily affect our view of approximately when the urbilaterian lived. The numerous and diverse Cambrian bilaterians suggests that the very first bilaterian lived *at least* as far back as the late Ediacaran. It seems unlikely that any one of the Ediacaran fossils that have been discovered actually *was* the urbilaterian (even if some of them were animals), but they may have been coterminous with it. Either way, we can bracket the likely time of existence of the urbilaterian to between about 600 and 550 MYA.

You might have noticed a complexity lurking in the above argument. Although there is much debate about exactly what the Ediacaran crea-tures were, there is no doubt that some of them were bilaterally sym-metrical – the picture of *Spriggina* that was shown in Figure 2.2 illustrates that point nicely. But if this and the other Ediacaran creatures were not animals, then their bilateral symmetry is an independent invention of this type of body layout. Such a situation would be broadly parallel with the independent, or convergent, invention of bilaterally symmetrical flowers by several different groups within the plant king-dom (Figure 8.4).

Determining *where* the urbilaterian lived is either harder or easier than determining when it lived; it depends on what's meant by 'where'. In using 'where', I am not referring to a particular place – that would be hard – but rather to a particular habitat type – which is much easier. Not only do we know that the urbilaterian must have been marine, because there was no terrestrial fauna until much later, but it was almost certainly a bottom-dweller, that is, an inhabitant of the sea floor. And we can be fairly confident that it was an animal that crawled over the sea floor rather than being affixed to it. Recall that sponges and sea anemones, which are attached to the substratum, are not bilaterally

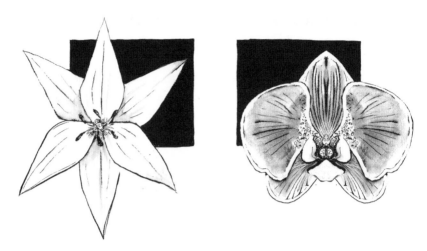

FIGURE 8.4 Different types of floral symmetry. Bilaterally symmetrical flowers, such as the one shown on the right, have arisen from radially symmetrical ones (left) several times in plant evolution. Bilateral symmetry of the body has probably arisen just once in the animal kingdom.

symmetrical. Bilateral symmetry almost certainly evolved alongside a mobile existence. If you're moving in a certain direction, it helps to have a head, with its concentration of sense organs, at the front.

The most important thing of all about the urbilaterian was not its exact chronology or ecology but rather its evolutionary potential. From it have arisen all the other bilaterian animals that we know of – in other words the vast majority of the animal kingdom. Given that bilateral symmetry and 'headedness' are closely associated, and that we humans are the ultimate species – at least so far – in evolutionary embellishment of the head, we owe a lot to our enigmatic urbilaterian ancestor.

9   Animal symmetry and heads

So far, we've encountered animals that are asymmetrical (sponges) and those that have radial symmetry (jellyfish). We've also dealt with the enigmatic urbilaterian – the first animal to have bilateral symmetry. We noted that most of the animal kingdom consists of bilaterally symmetrical animals, all of which have evolved from the urbilaterian that lived in Ediacaran times, probably between about 600 and 550 million years ago.

However, the issue of symmetry in animals is not as simple as a three-way split into lack of, radial, and bilateral symmetry. There are many fascinating nuances on the symmetry theme. To look at these we'll start with heads. In fact, where better to start than with our own heads.

If you look in a mirror you'll see a face that looks at first sight bilaterally symmetrical. Your eyes are about equidistant from your nose. Your nose itself seems like a symmetrical structure. The same can be said of your mouth. Of course your hairline may be asymmetrical due to a right-hand or left-hand parting. But that's fashion or habit, not biology, and such a parting can easily be shifted into the centre.

If you look more closely, though, either at your own face or at anyone else's, you'll notice that the symmetry is far from perfect (Figure 9.1). I don't think I know anyone whose nose is perfectly straight. If you were to make precise measurements on various facial features you'd find the same thing – very few of them are *perfectly* symmetrical. And there's no need to restrict our attention to the face. Try turning your hands palm side up and looking at the veins in your wrists that are carrying blood back to the heart. You'll notice that the pattern of spacing of these veins in one wrist is an approximate mirror-image of the pattern in the other wrist – but again close inspection or measurement will reveal that the symmetry is imperfect, and that the 'approximate' label is indeed justified.

This phenomenon, in which supposedly bilaterally symmetrical structures are not perfectly so, extends to all other bilaterian animals – there's nothing special about humans in this respect. Whether we are

FIGURE 9.1 Two human faces, both basically bilaterally symmetrical and yet both characterized by a degree of left-right asymmetry. The face on the left is the more symmetrical of the two.

dealing with dogs, birds, flies or frogs, measurements will reveal that their bilateral symmetry is imperfect.

Many studies have been undertaken in this area of departures from perfect symmetry. The existence of these departures, and the fact that they vary among individuals of the same species, whether human or otherwise, has been encapsulated in the unfortunate phrase *fluctuating asymmetry* (often abbreviated to FA). The reason I'm calling it unfortunate is that for most people 'fluctuating' refers to something changing in upward and downward directions over time, which is absolutely not what is being examined by students of FA. Rather, they typically examine differences between individuals within a species in their degree of departure from perfect bilateral symmetry; so the comparisons they are making are in space rather than in time.

Because of this unfortunate choice of phrase – a bad choice that got made a long time ago and then became embedded in the technical literature – I won't use it any more here. I had to mention it because there's a huge body of work on FA in the relevant scientific journals and books, and if you want to find it then the offending phrase that is often abbreviated to FA is a useful route in; but that's all, its usefulness extends no further.

While the phrase may be annoyingly inappropriate, the phenomenon itself is very interesting. There have been several findings in this area that appear to be general ones rather than species-specific ones – and these are always the more interesting to scientists, given that generalization is at the heart of the scientific endeavour.

One of these general findings is that, within a species, individuals that develop under more stressful conditions exhibit greater departures from perfect bilateral symmetry than those developing under less stressful ones. Care is needed here, though, because of the multiple uses of the word 'stress'. What is intended in the present context is physical rather than mental stress. An example should help to illustrate the kind of stress involved further.

Suppose that you are a biologist interested in this issue of departure from perfect bilateral symmetry. Suppose further that you want to do some rearing experiments to measure it and that, for ethical reasons, you decide to do these on insects rather than on mammals. So you rear flies of the same species at a range of temperatures and you use some simple measure of asymmetry such as the difference between the length of the left wing and the length of its right counterpart. The bigger the difference in these two measurements, the greater the asymmetry.

As you've probably guessed, this is not a 'just suppose' story at all. The experiment described has been done, with the following result. The further the rearing temperature is from the optimal temperature for the species concerned, the more asymmetrical are the flies. One way to think of it is that the fly's developmental system is trying to make a perfectly symmetrical product, but its ability to do so becomes progressively compromised, the more extreme, or stressful, the rearing temperature – in both hot and cold directions. Eventually, for any species, if we vary the temperature enough, there comes a temperature that is too hot (going upward) or too cold (going downward) for the developmental system to work at all – these two temperatures bracket the viable range for the species that is being studied.

Note that the characters discussed so far have been external ones: human faces and flies' wings. Although these show asymmetry, it is small in the grand scheme of things. To see this we need to delve inside a bilaterian animal; and again, what better animal to start with than the human one.

Internally, some of our organs are approximately bilaterally symmetrical, just as external ones are. For example, our left and right kidneys are approximate mirror-images, just as our left and right ears are. But our intestines are a tangled mess; we have a single, asymmetric heart; also a single liver and a single pancreas. So, while our departures from perfect bilateral symmetry on the outside are real, they are as nothing compared with their internal equivalents. What this all says is that calling an animal a bilaterian, or referring to it in a general way as being bilaterally symmetrical, is a kind of convenient shorthand. Most bilaterians have some features that are approximately bilaterally symmetrical, plus other features that come nowhere near.

It's always useful, in zoology, to mentally wander across different types of animal while considering any particular issue – including the symmetry issue we're currently focusing on, but many other issues too. This is because if we restrict our attention to one type of animal we may be tempted to see what we find there as representative of a large group of related animals, yet this may not be true.

As an example, we might reasonably think that what was said above about the symmetry or otherwise of human internal organs was true of vertebrates in general. But in many cases we'd be wrong. Consider, for example, the approximate mirror-image symmetry of our left and right lungs and kidneys. This approximate symmetry is as true of dogs as it is of humans, as any vet will be able to confirm. But what about snakes?

Most people, when they think about snakes, think of their heads with their (in some cases) poisonous fangs. If we think about their trunks, we picture them as rather featureless long tubular things, tapering to a point at the tail. Our thoughts rarely venture inside the trunk of a snake – but perhaps they should, because snake anatomy is very interesting and, from our own human perspective, very weird. Some snakes have only one lung; some others have two, but with one being tiny compared to the other.

Notice that the above account not only tells us that snakes are very different from humans in terms of their lungs, but also that they are different from each other. 'Humans' refers to just one species – it is plural only because there are many individual human beings. But

FIGURE 9.2 Dextral and sinistral shells of the same species of snail. The species is *Lymnaea peregra*, a common member of the freshwater snail fauna of the British Isles and parts of continental Europe. Most populations of this species have only the dextral form, but a few have both dextrals and sinistrals, the latter usually only present as a small minority.

'snakes' refers to about 3000 species. So if you were tending to think that this is a minor group of vertebrates, please think again.

Let's shift our attention from vertebrates to invertebrates, and specifically from snakes to snails. A typical snail looks anything but bilaterally symmetrical: it has a single, spiral shell that coils either to the left (sinistral) or to the right (dextral), when viewed from above. Right-coiling is commoner, but there is a substantial minority of species that coil to the left; and even a few in which the direction of coiling varies among individuals within the species (Figure 9.2).

The shell, of course, is a hard, essentially non-living structure. It's great for fossilizing, and so snails have a much better fossil record than most other animal groups. But it's not so great for informing us about the structure of the soft parts of the snail that are inside it.

To look at the animal denuded of its shell, we can, instead of trying to detach the shell (not easy, and not kind to the snail), look at those close relatives of snails, the slugs. Actually, some slugs have tiny, vestigial shells, sometimes on the outside of the body as expected, but sometimes internalized. However, other slugs have no shells at all – let's focus our attention on one of these. If you watch a large slug moving in a straight line across a path on a damp morning, you'll see a creature that looks bilaterally symmetrical. The long left-hand tentacle will be about the same length as the long right-hand one (there is usually a second pair of shorter tentacles, again symmetrical); and the two sides of the body itself will appear to be approximate mirror-images of each other, especially when viewed from a distance.

So, a perfect bilaterian? Well, actually, no. Both slugs and snails have asymmetric internal organs, and, connected with this, the openings of both their breathing systems and their reproductive systems (one apiece) are not in the midline, as ours are, but instead are on one side of the body or the other. And which side they are on varies. Recall those few species of snail whose shells can coil in either direction. It turns out that this variation is not something that is restricted to the shell. Comparing two snails of a variable-coiling species, we find that the respiratory and reproductive apertures switch sides along with the direction of shell coiling.

Now we come to the most bizarre bilaterians of all – the echinoderms. This name translates as spiny skin; the group includes the familiar starfish and sea urchins, plus other animals that are less familiar, such as the wonderfully named sea cucumbers. Most of today's echinoderms share an unusual kind of radial symmetry – pentaradial or five-fold. Let's focus on starfish. A typical starfish has five arms – albeit, as ever, nature is messy, which renders neat generalizations difficult. There are starfish with different numbers of arms, and while these numbers can be multiples of five they include numbers that are not – e.g. thirteen, a number of arms that is found both in some fossil starfish and in some extant ones (Figure 9.3).

Regardless of whether the number of arms is the typical five, or one of the less common numbers such as thirteen, it is clear that starfish, and other echinoderms too for that matter, are *not* bilaterally symmetrical. So why have I referred to them as bizarre bilaterians? Why use the word bilaterian at all in relation to these strange creatures?

To answer these questions we have to return to the issue of how animal groups are defined these days. In other words, we have to revisit a topic initially discussed in Chapter 7 on *How to make a tree*. There, we noted that it is current practice to use only what are called

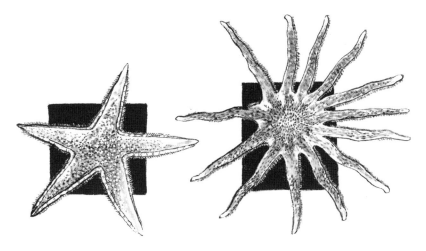

FIGURE 9.3 Two body layouts of starfish. On the left is the common five-arm arrangement. On the right is a much less common thirteen-arm pattern in which the typical pentaradial symmetry normally found in extant echinoderms has been lost.

monophyletic groups, namely those that include *all* the species that have descended from a given ancestral stem species – and no others. If we do that, then the group Bilateria must include echinoderms because they, like us, are descendants of the ancient urbilaterian. This is a bit similar to the vertebrate group Tetrapoda ('four legs') including legless snakes. Although at first this seems equally illogical – because snakes manifestly do not have four legs – enough is known about the evolutionary tree of vertebrates for the scientific community to be certain that snakes evolved from lizard-like ancestors.

The snake example, and many others, can be looked at in the following way: the name of the overall group – in this case Tetrapoda – is often devised because it describes a condition – in this case four-leggedness – that applies to the vast majority of its members. Some of its members, however, may have lost the condition concerned over the course of evolution. That's not really a problem. Taxonomically, snakes belong to the group Tetrapoda; anatomically, they are legless.

However, we should pause before concluding that the echinoderm example can be dealt with in the same way. No-one in their right mind would doubt that snakes are vertebrates and that they have evolved from ancestors with legs. But where do starfish and their kin belong in the animal evolutionary tree? Which non-echinoderm animals are they close to?

Back in the nineteenth century, some zoologists thought that the echinoderms were related to jellyfish and other cnidarians. So they created a group called Radiata and placed into it all animals that had radial symmetry – including the special case of pentaradial symmetry exhibited by extant echinoderms. Indeed, the recognition of the group Radiata precedes Darwin's theory of evolution by natural selection. Perhaps the first person to propose the group was the famous French comparative anatomist Georges Cuvier, who divided the animal kingdom into four *embranchements*, of which Radiata was one.

However, both comparative embryology and DNA sequence data have shown that the Radiata is not a natural grouping. So it is now clear that, just as snakes are tetrapods that have lost their legs, echinoderms are bilaterians that have lost their bilateral symmetry. Unlike jellyfish, echinoderms had bilaterian ancestors.

If this is true, the fossil record should help by revealing intermediates between the bilaterally symmetrical ancestors of echinoderms and the pentaradial symmetry that characterizes most of the extant forms. Also, considering the process of development in extant echinoderms, we might expect to find stages that are bilaterally symmetrical.

Both of these expectations are borne out. Some Cambrian fossil echinoderms exhibit what might be called 'quasi-bilateral' symmetry; and some larval stages of extant echinoderms are bilaterally symmetrical too. (There's more on this in Chapter 16.)

I said at the outset of this chapter that we would start with heads; let's also end with heads. Symmetry and heads are intimately connected because a head is the name we give to the anterior end of the main body axis of a bilaterian. If there is no bilateral symmetry, then, by definition, there is no head. If animal evolution had taken a different course than the one it did, and the animal evolutionary tree had consisted of a series of branches all of which led to irregular or radially symmetrical forms, with no branch leading to bilateral symmetry, then it seems likely that intelligence and consciousness would never have arisen. The Earth, as looked at by alien visitors, would then fall into the category of a planet with life but without intelligent life.

Of course, we should not equate heads with intelligence. To capture the relationship between these, it's useful to employ two criteria often used by mathematicians: *necessary* and *sufficient*. In my view, a head is a necessary prerequisite for intelligence but not a sufficient one. To put that another way, if you don't have a head you can't be intelligent; but having a head does not *guarantee* that you are intelligent.

This last point should not be interpreted as a jibe against some fellow humans. Intelligence does vary from one human to another, of course, just as do other characters, like height and weight. However, I'm referring here to much greater differences in intelligence than those that are found among humans. If we take a broad view of the animal kingdom, we see heads all over the place. Some heads may be such that the animals bearing them are not vastly less intelligent than us: chimp and dolphin heads come to mind. But other heads are much smaller, contain smaller and less sophisticated brains, and may convey on their bearers no more than a very limited repertoire of behaviour, and perhaps no thoughts – inasmuch as we can define these – at all.

You can see where this is going: to the heads of invertebrate bilaterians. Not all of these are rudimentary – the octopus, which will be a focus of our attention in Chapter 12, has an impressive head containing an equally impressive brain. But the majority of bilaterian invertebrates are much less impressive in this respect. Flies, which we discussed earlier, provide an example. But even lowlier than flies are the humble worms. These have heads, but not very big or brainy ones. Strangely, though, as we'll see in the next chapter, this has not prevented worms (there are many types of them) from being among the most ecologically successful animals on the planet.

10 A plethora of worms

In the introductory chapter, we looked at the structure of the animal kingdom in terms of the relative numbers of species belonging to different groups. There, we concentrated in particular on the relative numbers of vertebrate and invertebrate species; and we saw that vertebrates make up less than 5% of the animal kingdom. Here, we pursue the theme of the make-up of the animal kingdom further, but from a different perspective: not the proportion of it that is vertebrate but the proportion that is vermiform – or worm-like.

However, this issue is not as simple as it seems. There are many different ways of assessing the size of any particular group of animals. Also, as we have seen already, there are many different ways of defining 'group', especially in the light of Darwin's 'groups within groups' picture, for which he gave a mechanistic explanation – evolution. Prior to Darwin, the groups-within-groups view had probably been held by all serious students of zoology, and had been formalized by the Swede Carl Linnaeus – for plants as well as animals – way back in 1735.

Before looking at animal groups we'll look at animal sizes. By that I mean the actual body size of any animal – whether a worm or a horse. The conclusion we reached in Chapter 1, that vertebrates constitute only a few percent of the animal kingdom, was reached on the basis of species numbers alone. No other factors were taken into account. But some zoologists might say that this is inappropriate, and that body size should be taken into account too. If we adopt that approach, the contribution of vertebrates increases. This is because although size varies enormously *within* the vertebrates – contrast, for example, an elephant and a mouse – the average size of a vertebrate is clearly larger than that of an invertebrate. We can be certain of this, despite the existence of massive invertebrates like the giant squid, because such forms are vastly outnumbered by small creatures such as flies and worms.

This line of argument has brought us back to worms, and I should reiterate an important point here – that nature usually defies simple, neat generalizations. There are some worms that are more than 10 metres long. However, my main thrust in the present chapter is not to raise your opinion of worms by pointing to the existence of some very big ones – these are the exception rather than the rule. Instead, I want to raise the profile of the humble worm by using an argument based on groups rather than on body sizes.

Let's take a closer look at Darwin's 'groups within groups'. The biggest group that we're looking at in this book (well, after Chapter 2, anyway) is the kingdom Animalia. On the same level as that, but not our focus of attention here, is the kingdom Plantae. Notice that I have used the word 'level'. Any group of organisms, at any level, from kingdoms down to species, can be called a taxon (plural, taxa); and, related to this, the science of classifying organisms (both animals and others) is called *taxonomy*. Biologists like to think of taxa at the same level – it's sometimes referred to as having the same rank – as being in a sense equivalent to each other in scope. For example, the families Canidae (dogs) and Felidae (cats) are broadly equivalent, while the class Mammalia and the species *Homo sapiens* are not. (Such equivalence of families is a trickier concept across large phylogenetic distances, though.)

The kingdoms Animalia and Plantae provide an example of 'equivalence' at a high taxonomic level. At a much lower level, the species of bird called the greenfinch and the goldfinch provide an example, as do the tree species silver birch and sessile oak. There are often reasons to compare taxa at the same level with each other, but only rarely reasons to compare taxa at radically different levels. What would be the point in comparing a goldfinch with the entire plant kingdom?

To minimize jargon, I have generally avoided giving the names of levels of taxon intermediate between kingdoms and species. However, I have used three of them before – phyla, classes and families – and it is necessary to revisit one of these now: phyla. Recall (a) that this is the plural form – the singular is a phylum; and (b) that this is the next level down after kingdom, so it is still a very high-level taxon. We have discussed some individual phyla already – for example Echinodermata (starfish *et al.*) in the previous chapter and Cnidaria (jellyfish *et al.*) in Chapter 6 – but we have not asked the following, more general, question: how many phyla are there in the animal kingdom?

While not all zoologists agree on the answer to this question, because there is no universally accepted definition of a phylum, a typical answer would be 'about 35'. That number contrasts markedly with estimates of the number of known species of animal, which, as we saw earlier, are typically around 1.5 million. Such a contrast is expected, given the hierarchical groups-within-groups arrangement. There is only one species of human, but there are hundreds of primate species and thousands of mammal species; and our own phylum, Chordata, has more than 50,000 species.

Returning to the vermiform theme of this chapter, we can ask the question: How many of the 35 or so animal phyla can reasonably be labelled as worms? That's one of those deceptively simple questions, because it presupposes that we all know what we mean by 'worm' and, further, that we all mean the same thing. But this is not the case.

I have just fed my son's pet axolotl (a weird sort of salamander) with a waxworm. This creature is given its name because it has typical worm-like features: a cylindrical body tapering at each end; and no proper legs. But, far from being a worm, it is the larval stage of a moth – so it's an insect. There are many other examples of insect larvae being called worms. Hornworms are the larvae of a different kind of moth, while mealworms are beetle larvae.

All these usages can be ignored if we restrict the word *worm* to those animals that are worm-like at their adult stage. But there's another problem. If we want to answer the question of how many animal phyla are worm-phyla, we need to make a pragmatic decision along the following lines: we will define a worm-phylum as one in which the majority of its members are vermiform. That helps in several cases, notably in our own phylum Chordata and its main subphylum, the Vertebrata. Few people, either biologists or lay-folk, would consider the vertebrates to be worms. This is reasonable: most vertebrates that we see on a day-to-day basis, such as dogs, cats, cows or sheep, do not look worm-like at all. But there are other, less-frequently encountered, vertebrates that *are* worm-like and indeed have a corresponding name – for example the slow-worm (Figure 10.1).

We have already encountered two animal phyla most or all of whose members are worms: the flatworms and the segmented worms. But there are not just two worm-phyla. Rather, there are at least seven. Here are the common names of five that we have not yet met: round-worms, hair-worms, ribbon-worms, penis-worms and acorn worms. Together with the flatworms and the segmented worms, the fact that there are seven worm-phyla means that these collectively represent

FIGURE 10.1 A slow-worm. This animal is a vertebrate worm, in the sense of having an elongate, legless body. It is a member of the reptilian group called squamates (lizards and snakes). Snakes, slow-worms and various other legless groups evolved independently from lizard-like ancestors.

one-fifth of the 35 phyla in the animal kingdom (actually somewhat more than a fifth, because some minor worm-phyla are not included in the above list).

Luckily, this is not a conventional zoology textbook. If it were, you'd now be in for a long slog through the detailed structure, functioning and ecology of members of all the seven main worm-phyla. However, not being constrained by the textbook mould, I will feel free to briefly discuss a selection of interesting points about various different worms, and to leave those who want a more exhaustive and systematic account to consult an invertebrate zoology text such as the excellent one I mentioned before simply called *Invertebrates*, written by the Brusca brothers and published (2nd edition) in 2003.

Of the seven phyla of worms, I'm going to focus on two from now on in this chapter: penis-worms (phylum Priapulida) and roundworms (phylum Nematoda). The former could hardly be omitted, given their interesting name; they are also interesting as an example of a phylum that had many more species at an early stage in its history than it does now. The latter would be an even more major omission: they make up the biggest worm-phylum in the sense of having the most species. Very roughly, there are 25,000 known extant species of roundworms – in contrast to the approximately 25 species of penis-worms, a difference of three orders of magnitude.

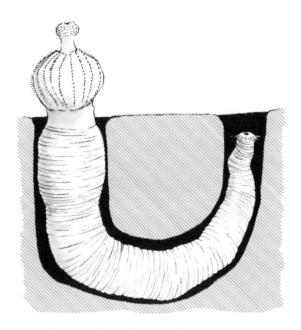

FIGURE 10.2 A penis-worm, shown living in a burrow in soft marine substratum. Fossil penis-worms from the Cambrian period, of which there are many good specimens, look remarkably similar to extant members of the group. Note that although the cuticle has a series of rings, so it is said to be annulated, the animal is not segmented in a more general way.

Let's look at the penis-worms first. They get their name from their appearance (Figure 10.2). The phylum is named after the Greek god Priapus, who had an enormous penis and was considered to be the god of all things related to reproduction: from harvests to animal husbandry to genitalia.

Penis-worms have a tough cuticle that is moulted as they grow. It is probably because of their possession of this cuticle that they have a rich fossil record extending back to the Cambrian period. The existence of so many fossils compared to some other worm-phyla is also a consequence of the phylum having been much larger in previous times than it is now. This possibility should always be remembered when considering the number of extant species in a phylum. Although today there are only a few species of penis-worms while there are very many species of roundworms, this does not mean that their relative numbers might not have been quite different – even reversed – during earlier periods of evolutionary history.

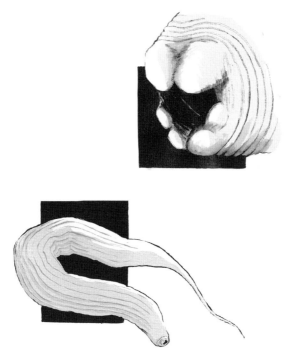

FIGURE 10.3 The model roundworm, *Caenorhabditis elegans*, with a close-up of the mouth. Both drawings are based on scanning electron microscope photographs, which are good at revealing surface features, but make the worm more solid-looking than it is when alive. Live worms have a translucent appearance. The length of an adult worm of this species is about 1 millimetre.

The fossil penis-worms from the Cambrian, dating from more than 500 million years ago, look remarkably similar to their present-day counterparts; and their ecology may have been similar too. Most extant penis-worms are marine predators, burrowing into soft, sandy or muddy sediments. They use a ring of spines around the mouth to capture prey. The diet depends on the size of the worm – the smallest are less than 1 millimetre long while the largest are more than 200 times that length.

And now, you might say, from the ridiculous to the sublime: from the bizarre penis-worms to an elegant roundworm: the species called *Caenorhabditis* (pronounced seen-oh-rab-dye-tiss) *elegans* (Figure 10.3).

Why start with – and indeed largely focus on – this particular species, given that there are about 25,000 kinds of roundworms? The reason is that it is one of a select few animals – less than 10 species out of more than a million – to have become what is called a model system for studying development. Recall that development is the relatively short-term process of making an adult from a fertilized egg, whereas evolution is the much longer-term process of turning one kind of animal into another – though, confusingly, if you read Darwin's 1859 *On the Origin of Species* or other literature from that time, 'evolution' then meant what we now call development and 'descent with modification' meant evolution.

Development, in the modern sense of the term, will be our focus of attention in Chapter 13. But we'll take a quick look at some aspects of it now with specific reference to the elegant worm, which became a model system in the 1960s, largely due to the efforts of South African biologist Sydney Brenner, who ended up winning a Nobel Prize.

The term *model system* might be self-explanatory to some readers but I'm guessing not to all. So here is a brief explanation. For a long time, in biomedical research, it has been common practice to use non-human mammals – often mice – as a sort of surrogate human. A possible new cancer treatment, for example, would be tested on mice before it is used in humans; so the mouse is acting as a biomedical model system.

In developmental biology, the rationale for having model systems is similar, but not identical, to the rationale that applies in medicine. Regarding how animals develop, a main aim is to understand the general mechanisms at work – such as key control genes switching other target genes on and off, and cells dividing in particular ways as the embryo grows. A lot of these mechanisms are thought to apply throughout the animal kingdom, and the goal is to reach a general, rather than a human-specific, understanding of them. So the developmental model systems are chosen not because they shed light on humans *per se*, but because they shed light on the *general principles* of animal development.

Therefore, rather than choosing a mammal such as a mouse, because the ethical issues surrounding experimentation are less severe than those that would be involved in experimenting on humans, the choice is based on ease of study and thus high probability of maximizing the acquisition of important results. As it happens, ethical difficulties end up being reduced too, because most of the developmental model systems are even less sentient than mice – inasmuch as we can ever really be sure about such things.

Anyhow, one key 'ease of study' characteristic to be sought in a developmental model system is rapid transit through a life-cycle or, to put it another way, a short egg-to-adult generation time. Until Brenner's advocacy of *C. elegans*, the small fruit-fly *Drosophila melanogaster*, with a generation time of only 2 weeks, in contrast to about 20 years in humans, was the main developmental model system. But Brenner's worm gets from egg to adult in about two days – the exact time depends on the temperature. So it's an even better model system for studying development. By the way, the terms *model system*, *model species*, *model animal* and *model organism* are all used by different authors in an essentially synonymous way.

There has been such an explosion of research, and experimental results, to have come out of the last half-century's work on Sydney Brenner's little worm that I cannot do it justice here. So I'll just pick a couple of important results and deal with those. One exemplifies commonalities in animal development, the other differences; we'll take them in that order.

I mentioned key control genes earlier. It turns out that many of these are common to most or all animals, despite the very different courses of development that are followed by, for example, worms, insects, molluscs and mammals; we'll discuss these genes further in Chapter 19. One particularly widespread group of control genes is the group called the 'Hox genes'. These were first discovered in fruit-flies. But later work on other model systems – including *C. elegans* – and indeed further work on non-model animals has revealed that this group of genes is found throughout the bilaterian animals and has a broadly similar function in all of them – patterning the main (antero-posterior) body axis as the animal develops. This is not the only function of Hox genes but it could perhaps be described as the primary one.

So the work on Hox genes in *C. elegans* and in other systems helped to reveal a commonality in the development of many animals. Now we turn to different research work on Brenner's worm which, in conjunction with related work on other systems, has helped to reveal an important kind of difference between the development of some animals and that of others. This difference concerns the degree to which the pattern of cell division that takes place during development is fixed or flexible, when we compare different individuals of the same species. If it's fixed, development is said to be *determinate*; if flexible, *indeterminate*. However, in the end these apparently neat categories have come to be seen simply as the opposite ends of a spectrum of degrees of determinacy.

The worm *C. elegans* is right at the determinate end of the spectrum. In contrast to most other animal species, where the number of cells in different adults can vary enormously, the number of cells in Brenner's worm is invariant. To be specific, the number is 959, with just four caveats. First, this is the cell number of an adult; obviously, earlier developmental stages have fewer cells. Second, it's the cell number of a hermaphrodite, which is by far the commonest type of adult in this species. Third, it's the cell number of the body as opposed to the sex cells, which are variable in number. Fourth, in a few of the worm's tissues, we need to count nuclei rather than cells, because in these tissues a cell membrane does not form around each nucleus in the normal way (tissues like this are called syncytial). But, putting these caveats to one side, if you take a sample of 100 worms of this species from their natural habitat – soil, where they eat bacteria – you will find that every single worm has exactly 959 cells, which is an astounding invariance in number.

To put this invariance in context, different adult humans differ in their total cell numbers by millions. Perhaps it's not a fair comparison because we are so much bigger than these tiny worms. However, if we were to pick an animal that is much closer to the worm's size than to ours – say a small insect – and repeat the exercise, we would typically find considerable variation in cell number from one adult member of a species to another.

It's not just the total number of cells that is fixed in Brenner's worm, but also the pattern of cell division that occurs to produce the 959 cells of the adult from a single cell in the course of development. This branching pattern cannot be shown here in its entirety because it's far too complex. But its beginning is shown in Figure 10.4. Note that some of the early embryonic cells contribute to just a single tissue-type in the adult, whereas others contribute to several tissues. This feature, like the others already described, is entirely constant – it characterizes all worms of the species.

Fascinating as *C. elegans* is as a developmental model system, it's good to recall that it's just one species out of about 25,000 in the roundworm phylum. What about the others? In terms of their development, the highly determinate system described above for *C. elegans* is also found in other members of the same family, the Rhabditidae. Indeed, it may be much more widespread than that – it may be typical

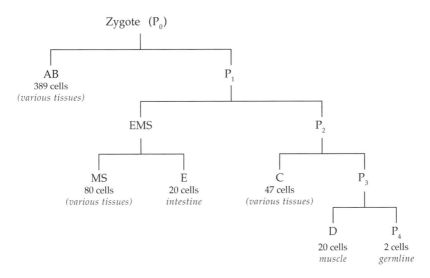

FIGURE 10.4 Pattern of early cell divisions in the development of the roundworm *Caenorhabditis elegans*. This pattern is consistent from one worm to another – in other words, it is fixed. The total number of cells indicated here, 558, applies to the hatchling. (Naming of cells: EMS stands for endomesoderm; this cell splits into E (endoderm) and MS (mesoderm). As can be seen, the E-cell gives rise to the intestine, which consists of endodermal tissue. P stands for posterior. A, B, C and D are just labels; the AB cell, which is shown here, gives rise to the A and B cells, which are not.)

of the whole class of which this family is a member (called the Chromadorea). However, this type of developmental system is not found in all nematodes.

In terms of their ecology, roundworms are found in nearly all types of habitat. There are marine, freshwater and terrestrial species. There are free-living and parasitic species. Some species live at incredibly high densities. One attempt to determine the density of roundworms in farmland produced an estimate of nearly 10 billion individuals per acre; in another, smaller-scale study, there were found to be almost 100,000 individuals in a single apple. So not only are there many species of roundworms, there are also astonishing numbers of individuals. Their combined ecological impact is colossal; and yet for most people this group of animals is virtually, or entirely, unknown.

11 Trends in animal complexity

A pervasive theme that has been with us since the beginning of the book – sometimes explicitly, other times implicitly – is that more complex animals are not necessarily more ecologically successful than simple ones. Indeed, this theme first emerged (Chapter 1) in a broader context than the animal kingdom – the context of life in general. The very first life-forms on Earth were probably rather like today's bacteria. The fact that bacteria and other unicellular forms continue to prosper attests to the fact that you don't have to be a big, complex organism to be fit – in the evolutionary sense of the word, as explained in Chapter 5. This point was reinforced in the previous chapter when we noted the incredible ecological success of roundworms, most of the 25,000 species of which are small and, in structural terms, quite simple, compared, for example, to arthropods or vertebrates.

It was this point (among others) that led me to state, at an early stage in the book, that evolution is not an escalator, up which creatures go at varying rates to higher levels of complexity. *Some* evolutionary lineages have shown rises in complexity over geological time, but others have not. Some have even shown decreases in complexity.

All this assumes that we can define complexity. Luckily, we can. In contrast to such subjective notions as the evolutionary 'advancedness' of an animal, its complexity can be measured in terms of its number of different types of component parts. Of course, nature being as it is, things are not quite that simple: parts exist at several levels. At the microscopic level there are cells; at the macroscopic level there are tissues and organs. But we can survive this problem because, in general, measuring complexity by using the number of types of part at one of these levels gives a similar result to measuring it at another level. Thus if animal A is seen as more complex than animal B when the number of different cell-types is used as our yardstick, animal A will probably also be seen as the more complex of the two if we count the number of different types of tissues or organs instead.

A problem does arise, however, when we descend to the molecular level and count the number of different genes. One surprise that has come out of the various genome studies of the last couple of decades is that the genomes of what seem to us to be very complex animals are not much bigger than those of simpler ones. A genome is the totality of the genetic material in a cell of an individual belonging to a particular species. There are various ways of measuring genome size, one of which is the number of different genes. The human genome, on which work began in the early 1990s, turned out to contain only about 25,000 protein-coding genes – about a quarter of the ball-park figure of 100,000 that many geneticists had been expecting. This is not a great deal larger than the 20,000 protein-coding genes in the genome of the little roundworm that we looked at in the previous chapter.

The reasonable correspondence of complexity comparisons between two kinds of animal when using number of types of cells or of organs is easy to understand. Since each organ requires different types of cell – those of a brain are very different from those of a heart – having more types of organs would seem to require having more types of cells. However, different cell-types do not possess different genes – rather, they have different genes switched on. This explains why the correspondence of complexity levels based on cell or organ types does not extend in a neat way to the level of the gene.

Perhaps more complex animals require not more genes but rather more complex patterns of gene-switching. In relation to this point, it is interesting to note a recent discovery made by American biologist Kevin Peterson and his colleagues. They studied a group of tiny molecules (called micro-RNAs) involved in gene-switching, and found that more complex animals had more families of micro-RNAs. Although this finding remains to be fully evaluated, it may provide a useful extension of the correspondence in complexity measures based on the number of types of parts down to the molecular level.

Anyhow, that's enough on measuring complexity. Now we need to look at how the level of complexity of animals has changed in the course of evolution. And this endeavour must be firmly anchored in the animal evolutionary tree. We've looked at trees before (especially in Chapter 7). But one aspect of them that we did not focus on then was what might be called their shape. This becomes our focus now.

Think of real trees for a moment – not animal evolutionary ones but individual plant ones, such as those in a forest. They are very variable in shape. Some conifers, such as spruce, have a quasi-triangular shape with a definite apex and straight, sloping sides. In contrast, a typical oak tree is much more irregular. Generally, though, plants that we call trees have a single trunk that has few branches near its base. Branching at the base is more characteristic of what we call a bush. On the other hand, there is no clear line of distinction between trees and bushes. Therefore, we can think of a continuum of bushiness. Although bushes are usually smaller than trees, let's ignore this, so to be bushier a plant doesn't have to be smaller, it just needs to have a lot of low-down branches sprouting in close proximity to each other so that it has no single trunk.

Now we can go back to animal evolution and ask: is the pattern of lineage splitting of any group of animals – or indeed of the animal kingdom as a whole – more like a tree or a bush? This is an important question because an evolutionary tree, especially if it takes the form of a conifer with a single apex, carries a certain philosophical baggage in terms of how we interpret it. Specifically, we feel a need to identify the animal that occupies the top position. In fact, this exercise is not an imaginary one; it has been conducted by many biologists using various tree pictures. And the animal at the apex of the tree is typically identified as being the human – us. Figure 11.1 shows a simplified, conifer-based version of this approach.

The trouble with this kind of picture of animal evolution is that it is too readily interpreted as implying that somehow the purpose of evolution is to make human beings. Other animals are side-branch distractions from the great upward thrust towards human-ness. This kind of view might be appealing to proponents of the non-scientific 'intelligent design' notion of evolution being micro-managed by a divine hand. But it has no basis in fact, because animal evolution is much bushier than suggested by the picture shown in Figure 11.1. For the more realistic animal evolutionary bush, see Figure 11.2.

It is hard to over-emphasize the difference between these two pictures and the ways in which they influence our thinking. The difference is a bit like that between Earth-centred (geocentric) and Sun-centred (heliocentric) models of our solar system. In the former, now discredited of course, our own planet was at the centre of things. In the latter, which we owe to Polish astronomer Nicolaus Copernicus, who published his revolutionary view in 1543, it is simply one of many planets orbiting the star that we call the Sun. Exactly how many depends on how 'planet' is defined. The modern definition excludes Pluto and so the number of

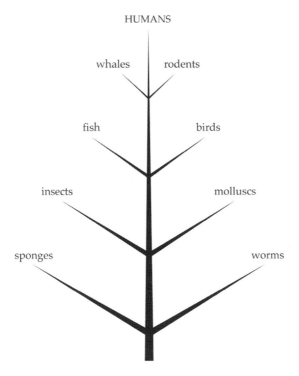

HUMANS

whales   rodents

fish   birds

insects   molluscs

sponges   worms

FIGURE 11.1 A conifer version of the animal evolutionary tree. Note the location of humans at the tree's apex. Such a depiction of evolution suggests that evolution has a goal – to make humans – and that other types of animal are somehow mere side-branches. Although depictions of this general form were often used in the past, today no biologists regard this as a valid view of evolution.

planets in the solar system is eight: Mercury, Venus, Earth, Mars, Jupiter, Saturn, Uranus and Neptune. This list, going from the closest-in planet to the furthest-out one, reveals our position as not even being special in the sense of coming first when counting in either direction.

Our lack of astronomical specialness extends beyond the solar system to our position in the galaxy. We are not in the galactic centre – lucky, because there is a super-massive black hole there. Rather, as noted in Chapter 2, we are part-way out along one of our galaxy's spiral arms.

Lack of evolutionary specialness is implied by the bush picture in Figure 11.2. And we should try to feel the full weight of this view of our place in the animal kingdom. The branching lineages embodied in this picture, and indeed their meandering nature, reveal evolution as something messy and unpredictable. Notice, though, that I have not called it

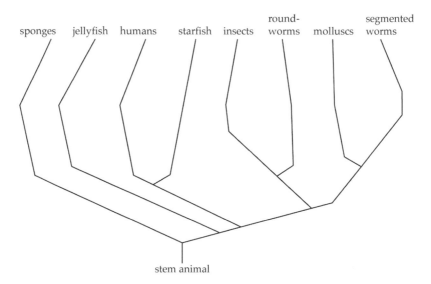

FIGURE 11.2 Animal evolution as a bush. Note the key difference between a bush and a tree: a bush has much early branching and thus no obvious trunk. The evolution of animals in general, and of animals belonging to particular groups, is more bush-like than tree-like.

random: that would be a bridge too far. Darwinian natural selection is a deterministic process, not a random one. It often results in particular lineages going in particular directions. But it has no foresight. Rather, within each lineage, it simply favours whatever form is fittest at a particular time and in a particular place. But if things change – ecological variables such as temperature shifting their values over time, or animals migrating and so moving from one place to another, where conditions may be different – then the form that was previously the fittest will probably not be the fittest any more. So evolution changes direction.

One of the main difficulties in picturing how complexity has changed during animal evolution is the limitation of the printed page to two dimensions. When a conifer tree is used to picture evolution with humans at the top, the vertical axis is effectively being used as some sort of uneasy hybrid between complexity and time. This doesn't work. A better approach is to deal with complexity issues in a time-independent way, as follows.

We can get rid of time by focusing on present-day animals. These all have an equally long period of evolutionary history, so in a tree diagram they would all occupy the same level on the vertical (time) axis. This

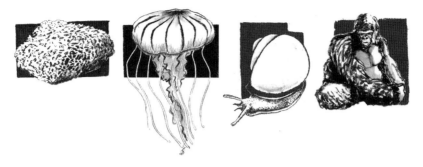

FIGURE 11.3 Four extant animals in a complexity series. The sponge is the least complex of all. The jellyfish is more so, with a wider range of cell-types, including the stinging cells that we looked at in Chapter 6. The snail is more complex again, with multiple internal organs. The gorilla, of course, is the most complex of the four, especially in relation to its nervous system.

fact allows us to ignore time altogether and to arrange extant animals in a series of ascending complexity. Of course, it makes sense to work with just a small number of extant animals because there are far too many to deal with at once. Let's take four kinds of animal that have been discussed in previous chapters – for diverse reasons – and arrange them in a complexity series (Figure 11.3).

No sane biologist would doubt the ordering shown in Figure 11.3. As we saw before, sponges have very few cell-types, and no nervous system. Jellyfish have a few more cell-types, including the stinging cell that was discussed in Chapter 6; also, they do have a nervous system, but no brain. Snails have multiple organs, including a rather minimalist brain. Gorillas have more organs than snails; their brains are large and consist of many different cell-types.

In Figure 11.3, not only time but also closeness of evolutionary relatedness was ignored. We can now bring back relatedness but without bringing back time too, by sticking with extant animals and making some interesting two-way comparisons in complexity. Again, it makes sense to focus on only a few animals. Here are just two very different comparisons, each pairwise, so together involving a total of four animals: humans and starfish; centipedes and flies.

The rationale for comparing humans (as representative vertebrates) and starfish (as representative echinoderms) is not yet apparent. It will become so in Chapter 16, but I need to pre-empt that chapter just a little here so that the choice of humans and starfish doesn't seem completely crazy. Our phylum and theirs are, despite all appearances, closely related, as relationships between phyla go. The rationale for comparing centipedes with flies is perhaps more obvious: they are members of two of the arthropod subphyla – Myriapoda and Insecta. So again, they are closely related, though again with the qualification that high-level groups such as phyla and subphyla are never *very* closely related.

Humans are clearly more complex than starfish. Starfish have taken a strange evolutionary route, as we noted in Chapter 9. Their ancestors were bilaterally symmetrical, but starfish, along with other extant echinoderms, are not. Along with losing their bilateral symmetry sometime in their distant evolutionary past, they lost the things that go with it – including heads and brains. Their nervous system consists of a simple nerve net. In contrast, the lineage that led to humans (and other mammals) underwent increases in both cell-type number and cephalization. So we have an unusually large number of cell-types (more than 200); and an unusually large brain, especially if looked at in relation to our body size.

Notice that although we are comparing two extant animals, time has already crept back into the argument. One lineage has experienced, over time, a decrease in complexity, the other an increase. Ultimately, evolution is all about changes happening over time, so we can never keep it out of the argument for very long. But our temporary banishment of it has enabled us to avoid conflating time and complexity; and to see that over time complexity can go up or down – or indeed stand still.

Turning to the second pairwise comparison – flies and centipedes – it's not too hard to see that the difference in complexity of the two forms is less than in the previous case. Both of these kinds of arthropod are bilaterally symmetrical and have brains – albeit rather simple ones. One approach to measuring complexity in arthropods, and thus of comparing any two or more of them, is to look at the number of different types of body-segments. Behind the head, a fly has several different types of segment. They can first be divided into thoracic and abdominal. Then they can be further subdivided, especially in the thorax, where the first segment has a pair of legs, the second has both legs and wings, and the third has both legs and small flight-balancing appendages called halteres (or, more casually, drumsticks).

Centipedes are very different in this respect. Immediately behind the head there is a single segment bearing the venom claws. But after

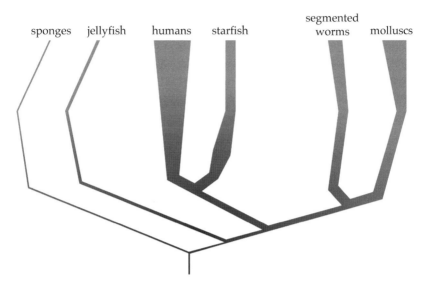

FIGURE II.4 An evolutionary bush with different branch widths. Here, the width of a branch represents the complexity of the animals involved. Note that, as time progresses, the complexity within a given lineage can increase, decrease or remain constant; also note that what is happening in different lineages over the same span of time may be different. (This tree has been slightly simplified in its number of branches compared with Figure 11.2.)

that there is no distinction between thorax and abdomen, and all the trunk segments bear a single pair of legs. So although there may be more segments in a centipede than in a fly, there are fewer *types* of segment. Hence the fly is the more complex of the two. But this time it is less clear if either lineage has undergone a decrease in complexity. Rather, it may be that since their last common ancestor (LCA) both lineages have increased in complexity, but the one leading to flies has increased more than the one leading to centipedes. We can't be sure of that, though, because identifying that crucial creature that was the LCA is far from easy. There were many types of arthropods in existence during the Cambrian period, more than 500 million years ago. And, interestingly, they were all marine. If we were able to trace fly and centipede lineages back with certainty to a particular Cambrian arthropod (we can't), we could examine all sorts of interesting things, not just structural changes (e.g. in segmentation) but also ecological changes – such as invasion of the land. Unlike vertebrates, which invaded the land just once (see Chapter 21), the arthropods invaded terrestrial habitats several times. Almost certainly, centipedes and flies are the result of separate such invasions.

Returning to the notion of an animal evolutionary bush, a possible way to include complexity in it, without conflating complexity and time, is shown in Figure 11.4. Here, the average complexity of a lineage is represented by the thickness of the line. This allows us to illustrate increases, decreases and standstills in complexity; also to show that in a single lineage these can all happen over different periods of its evolutionary history.

Although the view taken in this chapter has been that the big increase in complexity of the lineage leading to humans does not represent some sort of goal-seeking behaviour of evolution, but rather a kind of accident produced by the unpredictable meandering of evolutionary lineages in general, the take-home message should not be an entirely negative one. However we managed to rise to our current level of complexity, especially of our nervous system and brain, we are special because we have gone so far upwards in complexity that we have crossed a threshold that no other species has crossed: we can think, read and write about our own evolution. That's no mean feat.

12 Where the octopus is king

Always beware the use of 'king' for an animal. What do we mean when we invoke that old adage that the lion is the king of the beasts? Nothing very precise, for sure. Perhaps we mean that, as a ferocious predator, the lion is king of all the mammals of the African plains. It's a top predator: it eats many of them, but none of them eat it – at least when it is still alive. And what do we mean when we call one penguin a king, especially when there are emperor penguins too?

In calling the octopus a king in this chapter, I am not referring to its predatory habits, its size or its regal appearance, but rather to its brain. In a sense, the octopus is a king more like the human than the lion. But what is it king of, in this respect? Possibly of all the invertebrates; but I'm going to focus here on the realm of the molluscs. As you'll recall, the highest conventional level of group within the animal kingdom is the phylum (plural, phyla). We've already looked at several of these. Our own phylum, the Chordata, is mostly composed of the vertebrates, though it also includes a few close relatives that lack our characteristic spine. In Chapter 1, we saw that the vertebrates made up less than 5% of the animal kingdom, with their 50,000 or so species. The largest phylum by far in terms of species numbers is the Arthropoda, where the biggest constituent group (insects) represents, on its own, about three quarters of all animals.

We have also looked at other animal phyla that contain several thousand species. These include the sponges (8000) and the roundworms (25,000). A quick tour of these various figures might lead to the conclusion that although we vertebrates (and our chordate allies) are not the largest animal phylum – a distinction that goes to the arthropods – we are the second-largest such group. But in fact we come third. The molluscs, with nearly 100,000 species, are almost twice as speciose as us. They occupy marine, freshwater and terrestrial habitats; and they include forms that are very varied in size: from snails that are only about 1 millimetre in diameter to the giant squid, which can reach a length of more than 10 metres. In other words, their size range extends through four orders of magnitude.

The octopuses and squids belong to one subgroup within the phylum Mollusca. They are referred to as cephalopods (literally 'head-feet'). But despite their impressive size and intelligence, this is quite a small group of extant species – with fewer than 1000 of them recognized at present. So where do the other 99% of molluscan species belong? Well, about 70,000 of them are slugs and snails; about 20,000 of them are bivalves, such as cockles, mussels and oysters; and the rest belong to an assortment of minor groups. So, to understand the world in which the octopus is king we need to start with its simpler and more numerous cousins, and in particular the snails.

The difference between a snail and a slug lies in the possession or otherwise of a shell, as we saw in Chapter 9. A typical snail's shell coils in a particular spiral way – it's called an equiangular spiral – that results in the shell's aperture, from which the soft body of the snail protrudes when it is active and into which it can disappear when threatened, increasing in proportion to the overall size of the shell (Figure 12.1). Some snail-shells are different, perhaps most obviously so in the case of the limpet, whose shell is a simple cone. However, if we think of all snail-shells as tubes that widen as they grow, with some coiling and others not, then the difference doesn't seem as great as it did at first.

Most snails are marine, living in or by the sea. Familiar seashore snails include, in addition to limpets, winkles, whelks and cowries. These all breathe with gills. But when snails invaded terrestrial habitats, which they did about 300 million years ago (MYA), they evolved lungs.

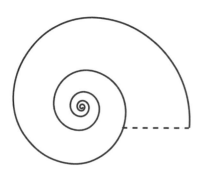 

FIGURE 12.1 An equiangular spiral (left) and a typical snail-shell viewed from above (right), which can be seen to conform, albeit imperfectly, to the equiangular design.

The bulk of today's land snails belong to a group called the pulmonates, a word that might sound vaguely familiar because of its similarity to the medical term pulmonary, referring to lungs. Strangely, though, freshwater snails that have evolved from these land snails have retained their lungs rather than re-evolving gills. There is a parallel here with vertebrate evolution. When land vertebrates evolved from fish they evolved lungs that replaced their gills. But when some lineages of mammals returned to the sea, notably the whales and dolphins, they also failed to re-evolve gills. Why this is so remains a mystery – which we'll examine in Chapter 21.

Invasion of the land not only involves the loss of gills but also, often, the loss of larvae. Marine snails typically develop from egg to juvenile via a tiny larva called a trochophore (Figure 12.2). This is a dispersal phase: these larvae float around in the sea. The larvae of coastal species eventually settle on a stretch of shoreline that may be quite distant from

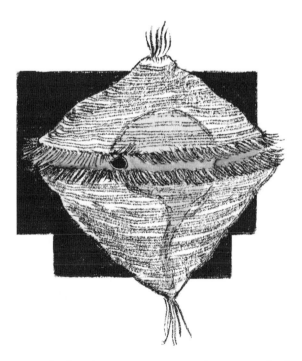

FIGURE 12.2 A trochophore larva. Most marine snails possess such a larva as a dispersive phase of their life-cycle. Some other groups of marine animals, notably polychaete worms, also possess this kind of larva. Indeed, the existence of a trochophore stage in the life-cycle has been used as a way of grouping related phyla.

that of their parents. Such a life-cycle enables expansion of the species' geographic range and avoids the over-population of any one bit of habitat. But on land this strategy doesn't work. Land snails lay their eggs in the soil; no small larva can happily float away from the laying site – except perhaps in a flood, which is too intermittent a phenomenon to be relied upon.

You can lose both your gills and your larva and still be called a snail. But the situation changes if you lose your shell. In that case, you are demoted – for that is how it is normally perceived – to being a slug. The typical slug has no shell at all, but, as mentioned earlier, some have tiny vestigial shells, often internal rather than being carried in the normal position outside the body.

The evolution of the slugs we're most familiar with – terrestrial ones – is something of an enigma. It's clear why a snail that has become terrestrial over the course of its evolutionary history should lose its gills and its larval stage. But why should it lose its shell? This impressive structure provides protection from both desiccation and predators. Surely in a terrestrial habitat, where the risk of desiccation is high, and predators are no less in evidence than in the sea, it would be sensible to retain your shell? This open question is no less mysterious than the earlier question of why freshwater snails evolving from terrestrial ancestors did not re-evolve gills.

However, the evolution of sea slugs is another matter. Unlike the much-maligned terrestrial slugs, sea slugs are very beautiful creatures (Figure 12.3). They often swim rather than crawl, and to call their elegant movement sluggish would be an insult indeed. Sea slugs (or nudibranchs, to use their official name) are able to swim as they do because the lack of a shell has allowed the body to evolve into a flatter form, often with what might be called 'wings' that flap to propel the animal through the water.

A caution is necessary at this point. Convergent evolution is a very common thing. So when I refer to evolutionary events in the singular, this is often a simplification of what has really happened. Not all sea slugs belong to the main group called the nudibranchs, because the reduction or loss of a shell has occurred more than once in sea snails. Equally, the invasion of the land by snails may have been a series of events rather than a single one – just as was the case with arthropods (but not, it seems, with vertebrates).

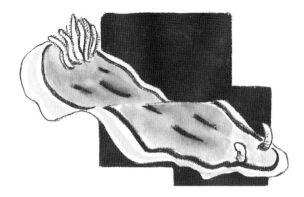

FIGURE 12.3 A nudibranch or sea slug (head with two tentacles to the right). As in the case of terrestrial slugs, the shell has been reduced – either to nothing or to a vestigial form. However, unlike terrestrial slugs, sea slugs are graceful creatures, swimming elegantly through the sea rather than sliding forward very slowly over land on a carpet of slime.

We have got this far in the present chapter without mentioning fossils. Now is a good time to bring them into the picture. Since most molluscs are snails, and since snails have shells, the fossil record of molluscs is excellent. Indeed, there are almost as many described and named fossil species of mollusc as living ones. There is probably no other animal phylum that can boast such a wealth of fossil species.

Just as trilobites are the most familiar fossil arthropods, ammonites are the most familiar fossil molluscs (Figure 12.4). Note the spiral form of the shell. This might be taken to indicate that ammonites were some kind of snails – but they were not. In fact, ammonites are more closely related to octopuses than to snails. Since octopuses either lack shells or have only small rudimentary ones, this statement of relationship requires some explanation.

As noted earlier, octopuses belong to the group of molluscs called cephalopods. This group also contains squids and cuttlefishes. (As an aside, note the plural forms of the last two. 'Cuttlefishes' implies several cuttlefish species, in contrast to several individuals of the same species; this usage is parallel to the use of 'fishes' for many fish species. Likewise, 'squids' is used to refer to multiple squid species.) Given that all three of these subgroups of cephalopods have many species, there is some variation within each in relation to shell form, but nevertheless a reasonable generalization is as follows. A cuttlefish has a flat internal shell; a squid has a pen-shaped internal shell; and an octopus, as noted earlier, has

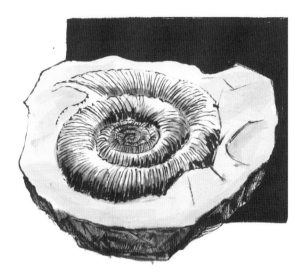

FIGURE 12.4 An ammonite fossil. Note that the shell is divided into many chambers, though fewer than the external rings might suggest. This is one of the main features that have led to an understanding of the relatedness of ammonites to various groups of extant molluscs. They are more closely related to octopuses than to snails.

either a rudimentary internal shell or no shell at all. The ancestors of these modern cephalopod forms had more impressive shells, so this is a case of evolutionary loss, not the first stages in the evolutionary origin of a novel structure.

Regardless of that last point, the spiral shell of an ammonite fossil resembles the spiral shell of a present-day snail far more closely than it does the pen-shaped shell of an extant squid. So what causes biologists to consider ammonites to be fossil cephalopods?

The answer to this question can be found in two places. First, note that the ammonite shell (Figure 12.4) is divided into a series of chambers. In this respect, it is very different from the shell of a snail, which is not (Figure 12.1). Second, there is a group of living cephalopods that I have not yet mentioned: nautiloids (Figure 12.5). These have the distinctive cephalopod arms, but their shell has not become vestigial or internalized. Rather, it is large and spiral, and it has chambers just as those of ammonite fossils do. The living animal occupies just the last (and hence largest) of these. This system works well, because the animal, its shell and the size of the last-formed chamber all increase with age. So when the animal is young it occupies a small chamber, and when it is adult it occupies a large one.

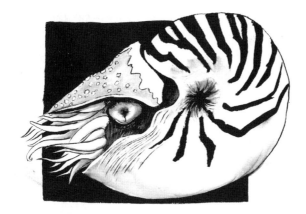

FIGURE 12.5 A living nautiloid; in particular, a species of the genus *Nautilus*. There are thought to be only about six remaining species of nautiloid. The group was much larger in earlier periods of geological time. It began in the Cambrian; over 2000 fossil species have been described.

There are subtle differences in the shells of ammonites and nautiloids that indicate that these two are not sister-groups within the Cephalopoda. But the two of them, plus the reduced-shell squids and cuttlefishes, and the rudimentary-shell or shell-less octopuses, are all related. So, a long time ago, cephalopods split from the rest of the molluscs and then split further (in a pattern that we needn't delve into the detail of). The first of these splits was at least as old as 400 MYA and almost certainly older, since there are ammonites of that age. In fact, ammonites persisted through a vast stretch of geological time, not going extinct until the dinosaurs did – about 65 MYA. The number of described species of ammonite is huge. This shows that while the Cephalopoda may be a relatively small group of molluscs today in terms of species numbers, it was undoubtedly much larger in the distant past.

Anyhow, recall that I have dubbed the octopus the king of the molluscs not because of the size of its group but rather because of its intelligence. Although this is renowned, we should dig into the evidence on which claims for octopus intelligence are based. In doing so we will find that some of it is only anecdotal. Luckily, though, some is based on careful observation and experiment. Together, the two types of evidence build a compelling case.

But we're not yet quite out of the wood of species numbers. So far I have said that there are nearly 1000 species of cephalopods. How many of these are octopuses? The broader of two possible answers to this question is 'about 300'. However, a narrower answer of 'about 10' can be given, if we include only those members of *Octopus* (note the italics and the initial capital, indicating a Linnaean genus name). In either case, we have a sizeable group, and thus considerable scope for variation in intelligence and other characters. Think about being asked: how intelligent is a typical great ape? This should be an easier question, given that the number of species of great apes is in single figures; but nevertheless it includes orang-utans and humans. As far as we know, orang-utans have not developed a theory of evolution (or relativity for that matter) while we clearly have done so.

The resolution of the problem of trying to determine the intelligence of a 'typical octopus' is based on the facts that (a) octopus species do not vary as much in intelligence as great ape species do, despite their greater number; and (b) most of the observational and experimental work has been done on relatively few species – one popular choice being the common octopus, *Octopus vulgaris*.

Finally, then, the big question: how do we know that octopuses are intelligent? And for that matter, what is intelligence? The answer to the latter question could be debated for ever. But here I will take the pragmatic approach that intelligence is inferred from certain types of behaviour, and in particular from learned, as opposed to instinctive, behaviour. Flexibility of behaviour, indicating a cumulative learning process in which the results of previous learning experiences can be refined by new ones, is especially important. Behaviour that indicates future planning, such as the accumulation of tools for use later on, is another important indicator of intelligence, as is exploratory behaviour, which, in the young of a species, might be described as play.

Many of the observations on octopuses that suggest intelligence come from aquaria. Lots of these are on record, though not all of them are in the scientific literature. Their absence from scholarly journals should make us wary but not necessarily dismissive: we should try to carefully evaluate each individual claim. An example of what not to believe is the story of 'Paul the Octopus' who, in 2010, apparently predicted the results of Germany's football matches in the World Cup. The predictions were made by seeing which of two boxes containing prey-items (mussels) Paul opened first. The boxes were marked with the flags of the relevant countries that were playing in each match. Although the reason for

WHERE THE OCTOPUS IS KING   121

Paul's apparent success is not clear (statistical fluke? human devious-
ness?), I cannot believe that it represents clairvoyance on Paul's part. I do
not believe in claims of human clairvoyance, so I see no reason to
believe similar claims made for an octopus.

Other stories about the antics of aquarium octopuses are more
believable. It seems that an octopus whose tank is not secure can escape
at night when no humans are present, climb into a fish tank, eat the fish
there, and return to its own tank before the early shift of staff appears the
next morning. Also, some octopuses seem to take likes and dislikes to
individuals, even when they are dressed the same (e.g. in aquarium
attendant uniforms). There are reports of an octopus repeatedly squirting
jets of water at disliked individuals but not at others.

There have also been observations on octopus behaviour with objects
that can be interpreted as tools. One study of this kind, published in
2009 by Australian zoologist Julian Finn and his colleagues, revealed
octopuses in their natural habitat carrying coconut shells and then using
them later to build shelters. Another study, this time laboratory-based,
showed that octopuses learned to play with empty pill bottles floating in
the water of their tanks. And another showed an octopus learning how
to undo the caps of other types of bottles. I have watched TV footage of
an octopus negotiating its way through an extremely complex maze in
order to find food. And octopuses use their arms to investigate foreign
objects protruding into their tanks.

There is a direct link from that last observation to the octopus
nervous system. Not only do these creatures have much bigger brains
than other molluscs, but their arms are very densely innervated. The
total number of nerve cells in the arms and brain has been estimated at
130 million. An octopus's brain is unusually large for an invertebrate –
it's about the size of a large walnut. That puts it firmly within the
vertebrate range of brain sizes.

Many people who have worked with octopuses have described looking
at these animals as an uncanny experience. One of the main reasons
given for this is a feeling that the octopus is looking back, and not just
that, but looking back at the person concerned in a 'what manner of
creature is this?' way. Of course, this is all very subjective stuff.

A digression is absolutely necessary here, when we are discussing
octopuses as looking at a person. The eyes they are using to look with
are incredibly similar to our eyes – indeed to vertebrate eyes in general.

This kind of eye is often called the camera-type eye, to distinguish it both from simple eyes (sometimes not much more than spots of light-sensitive pigment on the skin) and from the compound eyes of insects.

If you look at the layout of an octopus eye and a human one you will see all the same components arranged in a virtually identical way. The cornea, iris, pupil, lens and retina are all there in both. So is the main optic nerve running from the back of the eye to the brain. But the manner of connection between the individual nerve fibres taking information from the retina and the optic nerve is different. In human eyes, these individual nerve fibres bizarrely emerge at the *front* of the retina and so have to dive through it at a certain point, thus producing our blind spot, to get to the main optic nerve at the back. But in the octopus the individual nerve fibres emerge at the back of the retina – a simpler and better design with no blind spot.

A difference such as this between an otherwise near-identical pair of structures from different animals is a telltale sign of convergent evolution. The last common ancestor of octopuses and vertebrates lived more than 500 MYA in the Cambrian period or, probably, even earlier (the Ediacaran). Whatever it looked like, it did not have camera-type eyes. In fact, even within today's molluscs, cephalopods are unique in having eyes with this design. Snail eyes are comparatively simple. And bivalves, which I have barely touched on here despite their 20,000 species, are often eyeless.

So cephalopods are the kings of the molluscs in at least three ways. They have the greatest tactile ability, embodied in their multiple arms. They have the most complex eyes. And they have the biggest brains. These three attributes may well be related: complexity of tactile and visual information needs neural complexity to make sense of it. And neural complexity leads, ultimately, to intelligence, albeit this is a slippery and hard-to-define term.

Are cephalopods in general, and octopuses in particular, not just the most intelligent molluscs but also the most intelligent of all the invertebrates? We have still to look at some of the other major invertebrate groups, most notably the arthropods (in Chapter 14), so an answer to this question may be premature. Nevertheless, I'd put money on it being 'yes'.

13 How to make an animal

Every animal gets made by two processes, which take very different lengths of time. The longer-term process is the one we've already been discussing in most chapters of the book: evolution. Here, a particular type of animal is made from a different, earlier-arising type by a series of modifications that rely on Darwinian natural selection and perhaps, as we'll see later, on other things. The shorter-term process is the one we will now begin to address explicitly: development. Here, an animal is made from the starting point of (usually) a fertilized egg.

Although evolution and development work on very different time-scales, they are inextricably linked. Each is, in a manner of speaking, the starting point for the other. To see this clearly, it helps to consider the whole of egg-to-adult development as a *trajectory*, or, to put it another way, as a route from a simple, unicellular beginning to a complex, multicellular end. Each type of animal has such a trajectory, though when animals with very different adult forms are compared, their developmental trajectories are found to be likewise very different (especially in their later stages). For example, although we have not looked at any developmental details yet, it is clear, to use the molluscs of the last chapter as an example, that a very different route must be taken from the fertilized egg to end up in one case with a snail and in another case with an octopus.

Whatever ancient extinct mollusc was the last common ancestor of snails and octopuses, it must have had its own developmental trajectory – albeit we may never know exactly what this looked like because developmental stages fossilize much less frequently than adult ones. Nevertheless, we can be sure that the trajectory concerned, characterizing a very specific animal that lived perhaps 550 million years ago, was: (a) the result of prior evolution from even earlier forms; and (b) the starting point for future evolution that led, via different kinds of evolutionary modifications in different lineages, to the trajectories characterizing the snails and octopuses of today. To generalize from this example: the development of any animal is the result of past evolution (evolution begets development, if you like), but evolution can only produce new types of animal by modifying the developmental trajectories of previous ones (development begets evolution).

The relationship between evolution and development has been studied from the nineteenth century to the present. However, such studies can be divided into 'early' and 'recent', with a considerable period of time separating these. In the 1800s, there was what we now call *comparative embryology*, two leading proponents of which were Karl von Baer and Ernst Haeckel – more on these two later. In the period from about 1980 to the present, there is what we call *evolutionary developmental biology*, or evo-devo for short. We'll examine these interface disciplines – as well as the reason for the long gap between them – in later chapters. In the present chapter we'll concentrate on development itself, just leaving in the background for now development's relationship with evolution.

Way before Darwin, in the seventeenth and eighteenth centuries, there was a dispute about the nature of the developmental process between the proponents of two schools of thought: the preformationists and the epigenecists. The former thought that there was a preformed adult already present in the sperm (spermist preformation) or in the egg (ovist preformation) and that, during development, this miniature creature simply grew larger. The latter thought that there was no miniature adult present in either the sperm or the egg, but rather that these contained instead *information* which could be used to drive the developmental process. The debate was sometimes framed in the context of human development (Figure 13.1), but it can be applied equally to any other animal.

We now know, of course, that the epigenecists were right and the preformationists wrong – though anyone wanting a more detailed account of the history of this debate should read the 2004 book *Embryology, Epigenesis and Evolution* by the Canadian historian of biology Jason Scott Robert. There is no miniature adult in the head of a sperm, as depicted in Figure 13.1; neither is there one inside an egg. What is conveyed from one generation to the next is not a preformed animal but the information required to make one. Although the nature of this information was entirely unknown in the seventeenth century, it is known in considerable detail today. Much of the information is in the form of genes, and in particular their DNA sequences, which determine the kind of animal that the developmental process will build. However,

FIGURE 13.1 A miniature adult or homunculus inside the head of a human sperm. The preformationists of the seventeenth century thought that such homunculi existed, though some thought that they were to be found inside eggs rather than inside sperm cells. It is now clear that the idea of a homunculus was wrong. Both egg and sperm cells carry the information needed to make an adult, not miniature adults that will simply grow.

this is supplemented by crucial information carried by other molecules, such as those that will start to switch genes on and off as the process of going from fertilized egg to embryo begins.

It makes sense to examine how the process of development works in the temporal sequence in which developmental events occur. But that raises the question: is there a universal sequence that applies to all animals? The answer to this question is a predictable 'no'. However, perhaps surprisingly, there is a series of broadly defined stages that are common to most animals, and we can follow that series for the next while.

In most animals, development can be roughly divided into the following four processes, which occur in temporal sequence: cleavage, gastrulation, organogenesis and growth. Let's take these in turn.

Soon after fertilization, the egg – now called a zygote – begins to divide, or *cleave*. This latter word is used because the division is normally unaccompanied by growth. You start with a roundish object,

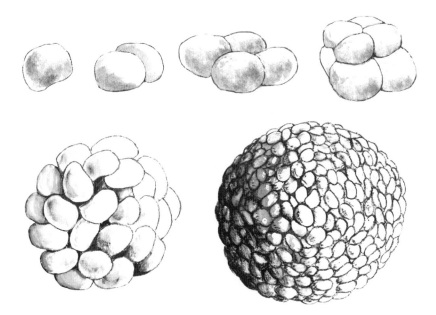

FIGURE 13.2 The earliest developmental process: cleavage. In most animal groups, the cells divide in a particular way. In some cases the top four cells of an eight-cell embryo lie directly above the bottom ones; this is called radial cleavage, which is the type shown here. In other cases the top four cells are rotated relative to the bottom ones; this is called spiral cleavage. The end-result of cleavage (bottom right) is a ball of many rather similar cells.

a single cell, and then cleavage furrows appear on its surface indicating division into first two, then four, then eight (and so on) cells which, between them, have approximately the same volume as the zygote had in the first place (Figure 13.2). In some animals the cells all look much alike at this stage. In others there is clearly visible variation in size. But in both cases the cells are all still what you might call unspecialized ones – they are not yet muscle cells, nerve cells or blood cells – such cell differentiation comes later. In most animals, the initial synchrony of division of the different cells breaks down, so that although early embryos can be classified into two-cell, four-cell, eight-cell and so on, at some point we no longer have embryos whose cell numbers are powers of two. For example, if asynchrony begins after the 64-cell stage, then there may never be an embryo with exactly 128 cells.

The spatial pattern of cleavage varies among different types of animal. In some, including humans, the four cells at the top of the embryo at the eight-cell stage lie directly over the four at the bottom (radial cleavage).

In others, for example molluscs, the top cells are rotated relative to the bottom ones (spiral cleavage). And in others, such as flies, cleavage starts by repeated division of just the nucleus, so that we end up with a single big 'cell' (it's called a syncytium) with lots of nuclei but no internal subdividing membranes – which only form later. So there are many variations on a theme. Yet these do not obscure the theme. Rather, they invest it with a certain diversity – exactly what we are coming to expect of the animal kingdom, with its interesting combination of unity and variety.

The end-result of cleavage is a ball of cells – usually a hollow one with a central cavity that may be filled with fluid or yolk, and sometimes with an inner group of cells. This ball goes by different names in different animal groups – but generally the names begin with *blast* (from the Greek *blastos*, for 'sprout'). So we have the amphibian blastula, the mammalian blastocyst, the avian blastodisc and the insect blastoderm, for example. The important thing to remember is the general structure of this developmental stage (hollow ball of cells, with the ball being sometimes spherical, sometimes ovoid like a rugby ball, and sometimes flattened as in a deflated and squashed beach ball), rather than the names.

The second stage in the development of almost all animals – called *gastrulation* – is characterized by cell reorganization more than by division. To understand it, we need to consider the question of what constitutes the top and bottom of the earliest embryonic stage. The top of a blastula is called the animal pole, the bottom the vegetal pole. This is a slightly odd pair of terms, because of course vegetal sounds like vegetable, but clearly no part of an animal embryo develops into a plant. It is sometimes true, though, that most or all of what will become the animal does start off in the animal pole of the blastula. This is because, in many yolk-rich eggs, the vegetal pole is mainly a yolk-containing region. But, as always, we need to be careful to acknowledge nature's variety. In many insect embryos, the yolk is concentrated in the middle.

The important thing here is to emphasize what the poles of an early embryo are *not*, rather than what they are. They are *not* the rudimentary head and tail ends of the animal that will follow; or, to put it another way, they do not correspond to the ends of the main (anterior–posterior) body axis. This axis only arises through gastrulation – another odd term with a Greek root, this time *gaster*, for 'stomach'. The reason for the connection with stomachs will become apparent shortly.

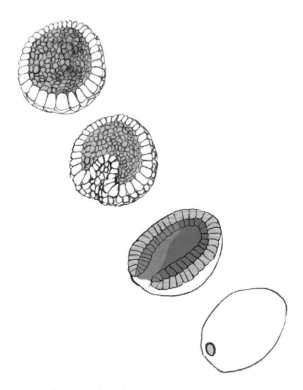

FIGURE 13.3 Gastrulation and its three main results: the antero-posterior body axis, the gut tube, and the production of the three germ layers referred to as the endoderm, mesoderm and ectoderm (shown by different shading here, with the endoderm darkest). The inward movement of tissue that is involved in gastrulation has been likened to someone poking a finger into a balloon: see text.

A good way to think about the embryonic contortions that constitute gastrulation is to imagine the early embryo as an inflated balloon and the process of gastrulation as someone poking a finger into the balloon at some point around its periphery. Assuming that our imaginary balloon is tough and not prone to bursting, the finger can be pushed a long way inside (Figure 13.3). Eventually, in the embryo (but not in a balloon), the tunnel thus made can, and often does, emerge at the other side. Given that the embryo is elongating as this happens, we end up with a sausage-like form with a tube through it. (Exactly how the gut tube is made varies among different groups of animals; more on this in Chapter 16.)

At this point gastrulation has ended and it has produced three things. First, it has produced the main body axis, whose ends do now correspond with the equivalent (antero-posterior) axis in the larva or adult. Second, it has produced the gut – the tube running from mouth to anus (hence the link with stomach). Third, although I haven't mentioned it yet, the contortions described above are accompanied by cells becoming committed to belonging to one of three layers of embryonic tissue, called endoderm, mesoderm and ectoderm (literally inside, middle, and outside skin). These will have different fates as development proceeds. For example, gut tissue is derived from endoderm, muscle tissue from mesoderm and neural tissue from ectoderm.

Before proceeding to look at later stages of development, it is again necessary to emphasize that no simple scheme, such as the one above, can be used for the whole animal kingdom, given its immense variety. At the very least, we should note that aspects of Figure 13.3 do not apply outside the bilaterian animals, because in asymmetric and radially symmetric animals there is no head-to-tail axis and no equivalent of our mouth-to-anus gut. Recall that in a jellyfish the same opening functions for food ingress and waste egress. Also, jellyfish and their allies probably do not have three embryonic tissue layers (there is some debate on this), and sponges certainly do not have three such layers.

With that caveat we can continue. The next stage in animal development is *organogenesis* – a reasonably self-explanatory word. Here, the organs that the animal will need to render it an autonomous and fully functioning organism are made. In vertebrates, the central nervous system is often the first organ system to begin to form, and this process is given its own name – neurulation. But in many invertebrates it makes no sense to single out the nervous system – or any other system – and to name it separately. So here we'll be content to use organogenesis to cover the formation of *all* the organs, whether brain, heart or something else.

It's clear that exactly what happens in the formation of one organ must be different to what happens in the formation of another. So, at this point in our temporal tour of development, we need to look at the general principles involved rather than lots of organ-specific details, in which it would be all too easy to drown. As an organ forms, two main things happen. These can be referred to as within-cell things and between-cell things; or, if you prefer, as (roughly) microscopic things and macroscopic things. We'll deal with these in turn.

The main thing that happens within each cell is a transition from a generalized to a specialized state: this is the process of cell differentiation, through which one cell becomes a nerve cell, another becomes a muscle cell, and so on. This process could conceivably happen in two ways: either (a) the genes that are unnecessary to make (say) a muscle cell could be eliminated from a cell that is destined to contribute to a block of muscle tissue, or (b) the unnecessary genes could be switched off. The former mechanism would be irreversible, the latter at least potentially reversible.

The classic experiments which proved that the second mechanism was the basis for cell differentiation in animals were done by the British developmental biologist John Gurdon and his colleagues at Oxford, and were published in a series of papers, the first in 1958 – over half a century ago. Gurdon was awarded a Nobel Prize for this work in 2012. The experiments involved destroying the nucleus of a frog's egg and then injecting into it the nucleus of a differentiated cell from later in the development of another frog (or tadpole) of the same species. If all the genes needed to make a frog were still present in this nucleus, with some of them switched off, there was at least a chance that they could be switched back on again by agents in the egg cytoplasm. Amazingly, after many trials, Gurdon and colleagues were able to demonstrate this happening: they managed to produce tadpoles and frogs from their engineered hybrid cells.

So the frog was the first animal to be artificially cloned. This is still news to many people, who think that the first such animal was the famous 'Dolly the Sheep', produced in much the same way as described above, in a laboratory just outside Edinburgh, by Ian Wilmut and colleagues. This confusion is probably another manifestation of the animal/ mammal mix-up that I mentioned at the start of the book. Dolly was indeed the first mammal to be cloned (in the 1990s); but she was most certainly not the first cloned animal.

Now we can turn to the between-cell aspect of organogenesis. If each individual muscle is considered as an organ, then one thing that strikes us straight away is that these organs differ a lot in shape. Your biceps is spindle-shaped. In contrast, there are muscles running across your shoulder blade that are essentially flat (Figure 13.4). You would have a strange appearance if these latter muscles, and also the flat muscles of your forehead, were shaped like spindles. We often take it for granted that each muscle ends up being the right shape for its position in the body; but we should not, because some mechanism is

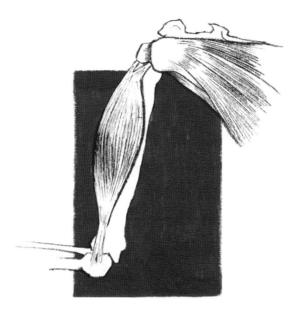

FIGURE 13.4 The shapes of different muscles in a human, illustrated by the biceps muscle of the upper arm, which is spindle-shaped, and the muscles that span the shoulder blade, which are flattish.

needed to ensure that this happens. And it must be a different mechanism from the one underlying cell differentiation, because the cells of each muscle are the same. It is not their internal structure that differs between, for example, biceps and forehead muscles, but rather their arrangement in space.

The achievement of the correct arrangement of cells, and hence the correct shape and size of organ, is referred to as morphogenesis. This process depends on at least three things happening at the level of the constituent cells. The main thing is that cells must divide at a certain rate, in certain planes, and for a certain duration; then they must stop dividing at the right time. In addition, cell movement may be involved. So also may be some programmed cell death – this is called apoptosis. The reason why ducks have webbed feet and we don't is that there is more cell death in our interdigital regions than in theirs. It's also important that cell differentiation is coordinated in space and time – this is sometimes considered as a separate process called pattern formation, though it could equally be considered as part of morphogenesis, broadly defined.

After organogenesis, the fourth and final – with some caveats – phase of development is *growth* of the animal with its now-formed organs. The growth of an animal can take two forms, called isometric and allometric. In the former, everything grows at the same rate, so the bigger, older animal that grows from its smaller, younger counterpart is scaled up in size but identical in shape. In the latter – which is the norm – different parts grow at different rates, with the result that shape changes with growth. A well-known example is the decline in the size of the human head, as a proportion of total body size, during development from late embryo through infancy to adulthood.

So what are those caveats mentioned above? Well, nothing in biology is neat and tidy. There is not, in any animal, such a clear distinction between organogenesis and later allometric growth as previously implied; rather, one gradually gives way to the other. Also, morpho-genetic processes can restart a long way into the growth phase, as evidenced, in human development, by the changes that occur during puberty, and by the strange appearance, after sexual maturity, of an extra four teeth – our so-called wisdom teeth.

In other animals, the caveats are much greater. The development of a butterfly starts with cleavage, gastrulation, organogenesis and allometric growth, just as human development does. But the result of all those developmental processes is a caterpillar. The progress of this to becoming a butterfly involves a series of moults and then, most incred-ibly of all, the process of metamorphosis that takes place in the chrysalis or pupa. Here, most of the caterpillar's tissues and organs are broken down and replaced with new ones that will characterize the butterfly. The new structures are made from tiny, inconspicuous disc-shaped bits of tissue that the caterpillar was carrying inside itself while going about its business of eating in order to grow – it has to reach a certain minimum size for metamorphosis to be possible.

Developmental systems without metamorphosis are referred to as direct; and the animals possessing them are said to have a simple life-cycle. In contrast, systems with metamorphosis are referred to as indirect; and the corresponding life-cycles are said to be complex. This is a huge difference. But there is also much variation within each of the two categories. Most frogs have indirect development via a tadpole stage that metamorphoses into an adult – but there is little similarity in detail between their metamorphoses and those of butterflies. Equally, in the realm of direct development, there is little in common between the development of (say) a centipede and a gorilla: both have direct development, and thus lack a larval stage and a metamorphosis, but other than that the commonalities at the organismic level are few. (There are more commonalities at the molecular level, but that's another story: see Chapter 19).

The overriding message here is that different animals have both commonalities and differences in their development. Given the existence of the latter, it matters a lot which animals we study: our understanding of development will be much influenced by the choice of these. At present, there is a drive to spread out and study the development of an ever-wider range of animals, representing many of the phyla of the animal kingdom. This move is to be applauded, though it is still in its infancy. Most of the information that we have about development, especially the information generated before about 1980, when the spreading-out initiative began, comes from a rather small number of animal species – as noted earlier, these have come to be known as model systems. The big six model animals include four vertebrates and two invertebrates, as follows. Vertebrates: mouse, chick, zebrafish and frog (*Xenopus laevis*; there are about 6000 species of frog, some of which lack a tadpole stage). Invertebrates: the fruit-fly *Drosophila melanogaster* and the soil-dwelling roundworm *Caenorhabditis elegans*, the highly determinate nature of whose development we noted in Chapter 10.

Of these, the first to be seriously studied was the chick, mostly due to the large size of hens' eggs – compared, for example, with those of a mouse. Of course, the hard, non-transparent shell poses something of a problem, but a careful student of chick embryogenesis can cut away sections of egg shell, thereby producing, literally, a window through which to watch development in progress. Alternatively, eggs can be broken open entirely and their contents observed, at various stages of development. These methods were used by the Italian anatomist Hieronymus Fabricius (1537–1619), who has been described as the father of embryology. His work *On the Formation of Eggs and Chicks* was published posthumously in 1621.

We have come a long way since Fabricius's descriptive embryological studies of the late sixteenth and early seventeenth centuries. Current developmental biology concentrates on the mechanisms underlying the observable phenomena that the early embryologists revealed. The nuclear transplantation experiments of the mid-to-late twentieth century, as described above, were a far cry from, and yet relied upon, the early descriptive work. And in the twenty-first century, most developmental biology involves dissecting development at the level of molecules – DNA, proteins and others. We'll examine the results of such studies in Chapter 19.

14 Exoskeletons galore

In this chapter, we examine the nature of those animals that collectively form the biggest group in the animal kingdom – Arthropoda – and explore their evolutionary relationships, both with each other and with other animals. The Arthropoda is the most species-rich phylum by a very long way. There are well over a million species of arthropods (most of them insects); this means that they make up more than three-quarters of the known animal kingdom. Arthropod species inhabit almost all environments on Earth, from the deepest oceans to the driest deserts.

The key feature underlying the success of the arthropods is the exoskeleton. In fact, a hard structure acting as, among other things, an anchor-point for muscles, is a key feature in the second- and third-largest groups of animals too. Most molluscs have a shell, or a pair or series of them, as we noted in Chapter 12; and species of our own group, the vertebrates, are characterized by possession of endoskeletons. Not only do all these hard structures provide a key part of a crucial biomechanical system that allows movement, but they also allow more frequent fossilization than do soft tissues.

Before we get into any more detail about the arthropods, we should deal with a question, which emerges from the above argument, about animals that lack hard parts – and in particular the many kinds of worms discussed in earlier chapters, especially Chapter 10. How do muscles function in such animals if they have no hard parts to attach to? This question might equally well be asked of the familiar earthworm, which can grow to about 10 centimetres, or of a tiny roundworm, which, at a mere 1 millimetre, is two orders of magnitude smaller. In general, how can the muscles that such creatures have actually work at all if they have no skeleton or shell to attach to?

A common arrangement of muscles in a vermiform animal consists of two layers: circular muscles and longitudinal ones. By various patterns of contraction, these can produce forward and backward movement of the animal. Of course, if the animal lacks hard parts it must change shape as muscles contract: watch an earthworm that has foolishly come to the surface on a wet day as it moves over paving stones and you will see such shape changes happening. So a workable system of muscle-powered movement without a hard skeleton is possible, but it can easily be outdone.

Consider, in contrast, the lightning-quick movement of the insect known as a praying mantis (Figure 14.1) as it catches an item of prey. Here, muscles are firmly anchored at many points on different parts of the exoskeleton, both in sections of the 'arms', which are really its forelegs, and in sections of the trunk. Their action is complex, rapid

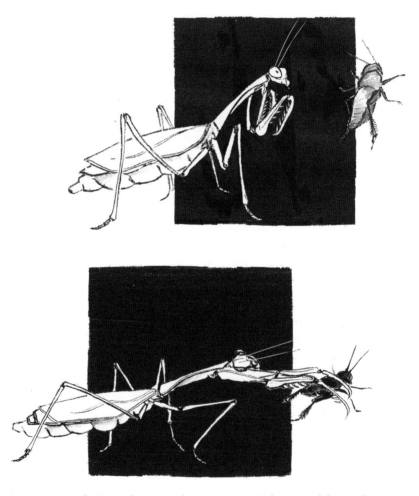

FIGURE 14.1 The insect known as the praying mantis because of the similarity between the way it hold its 'hands' and the way in which praying humans hold theirs. The mantis is shown here in two positions: prior to, and after, the capture of a prey item.

and precise. By the way, that's not a spelling error: it really is called a *praying* mantis (one of the species has the Latin name *Mantis religiosa*), despite the fact that it's a predator that spends a lot of its time *preying* but, as far as we are aware, no time at all praying. Its name derives from the superficial similarity between the way it often holds its forelegs and the way humans often hold their hands when in prayer.

Of course, we can't afford to take a species-by-species approach to the arthropods: there are far too many of them for that to be a sensible strategy. Rather, we need to examine them in large groups, with just occasional mention of individual species to illustrate some specific points. So, of what large groups are (and were) the arthropods composed?

The extant forms are divided among four subphyla, but the distribution of species across these is very uneven. There are about a million species of insects; about 100,000 species in the Chelicerata (spiders, scorpions, mites, ticks and related forms); about 70,000 in the Crustacea (lobsters, crabs, shrimps and their allies) and about 15,000 in the Myriapoda (centipedes and millipedes). The extinct arthropods that don't fall into any of these four groups are mostly trilobites. But there are also others, such as the anomalocarids that we looked at when discussing the Cambrian explosion in Chapter 4.

Arthropods, like most animal phyla, began in the sea. There are plentiful arthropod fossils from the Cambrian period, indicating that there had already been much evolutionary diversification of these animals by 500 million years ago (MYA). But if we look at the extant forms, three out of the four subphyla consist largely of terrestrial creatures. Insects are mostly terrestrial, though many have freshwater larvae. Arachnids and other chelicerates are almost entirely terrestrial. And there is not a single truly aquatic species of myriapod.

In terms of habitat, then, the crustaceans most closely resemble the Cambrian arthropods. Although a few crustacean lineages have invaded the land, with the best-known of these being the woodlice, the number of terrestrial crustacean species is a tiny fraction of the overall figure of about 70,000 species for the group as a whole. The vast majority of crustaceans are aquatic, and most of these are marine – for example, lobsters, crabs and barnacles. Most of these creatures breathe, as you might expect, with gills; though some very small aquatic crustaceans with thin cuticles use their entire body surface to breathe. Gills are defined as structures that allow the uptake of oxygen from water as opposed to from the air. The exact form that gills take varies enormously, both between distantly related animal groups (e.g. fish and crustaceans) and between more closely related groups, such as between some crustaceans and others. Many crustaceans have their gills in some of their appendages, though exactly which appendages varies; others have gills in chambers that are located inside the carapace (the exoskeletal head shield that is particularly obvious in crabs).

Although the terrestrial woodlice are familiar animals, it's not immediately clear from observing them how they breathe. In fact, they have air sacs in their rear pairs of legs – these are in effect lungs, because they allow oxygen uptake from the air.

It's interesting to compare the air-breathing systems of woodlice with their equivalents in other types of terrestrial arthropods. Spiders and their kin use structures called book-lungs. The name comes from the book-like appearance of this type of lung: there are many separate layers (pages) – this is an arrangement that serves to maximize the surface area for gas exchange. Such maximization is a common requirement of most types of breathing apparatus – including our own lungs, where it is achieved in a very different way. Spider book-lungs are located in the trunk (not in appendages) just behind what might be called the waist. I'm using this term in a casual way – you won't see 'waist' in any arachnology textbook – to refer to the narrow region of trunk that joins the head end of a spider with the tail end; the official names for these two parts of the body are the prosoma (front body) and the opisthosoma (rear body).

In contrast to the leg-borne air sacs of woodlice and the rear-body book-lungs of spiders, both insects and myriapods have systems of air tubes called tracheae ramifying through the body. These open to the exterior through holes called spiracles. The number and location of spiracles varies. For example, in centipedes the spiracles are paired and lateral in most groups but unpaired and located in the dorsal midline in one group.

You might well ask where this argument is going. The answer is that the comparative information above is one of many types of such information that have been used to try to discern the pattern of relationships among the four major groups of arthropods. This information has also been used to address the related issues of: (a) the number of independent arthropod invasions of the land; and (b) the frequency of convergent evolution of particular air-breathing (and other) structures.

The facts so far would suggest that woodlice, spiders, and insects/ myriapods invaded the land separately. However, the use of 'insects/ myriapods' is deliberately hinting at the hypothesis that this combined group might have had a last common ancestor (LCA) that was terrestrial, given that both insects and myriapods have tracheal systems. As we'll see shortly, this hypothesis is almost certainly wrong.

Let's turn our attention to another feature of arthropods: their development and, more generally, their life-cycles. This will provide a link with the previous chapter, which was essentially an overview of animal development.

Indirect development, via larvae, is commonplace in the arthropods. However, the details vary enormously from group to group. Many marine crustaceans have small, short-lived larvae, which can be thought of as the dispersive phase of the life-cycle. A common type among several crustacean groups is called a nauplius larva (Figure 14.2). In contrast, the larval stages of insects are typically much larger and longer-lived; they are a growth phase rather than a dispersive phase of the life-cycle, spending a good deal of their time feeding. This is true

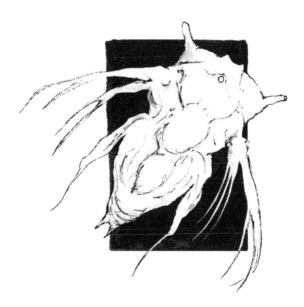

FIGURE 14.2 The developmental stage referred to as a nauplius larva that is part of the life-cycle of many marine crustaceans. Like many other kinds of marine larva, this one has a dispersive function.

both of freshwater larvae (e.g. dragonfly nymphs) and of terrestrial larvae (e.g. lepidopteran caterpillars).

The myriapods generally have direct development, so there is no true larval stage. Nevertheless, their young can be noticeably different from the corresponding adults in features additional to the obvious one of overall body size. Interestingly, some centipedes hatch from their eggs with the full adult complement of body-segments (even though this can be as high as 191), while other centipedes, and all millipedes, hatch with fewer segments than are present in the adult – in which case they must add segments as they grow. This is done via a series of moults, in each of which one or more segments are added at the posterior end of the body.

Most millipedes hatch with three pairs of legs. This suggests a link with the insects, which have three pairs of legs as adults (and in some cases as juveniles too). This fact used to be added to the similarity of the insect and myriapod tracheal systems to reinforce the hypothesis that these two are sister-groups. However, that hypothesis has not been supported by the more recently acquired comparative molecular data, in the form of DNA and protein sequences, almost all of which point to a sister-group relationship between the insects and the crustaceans. Indeed, a new group name has been invented to cover the insects and crustaceans jointly – the Pancrustacea.

Assuming that this new hypothesis is correct, how do the other two arthropod groups connect, in an evolutionary tree, with the combined group of insects and crustaceans? The predominant view is shown in Figure 14.3. Here, Myriapoda is the sister-group to the Pancrustacea, with the other group of extant arthropods, the Chelicerata (arachnids and a few others), being the outgroup to the rest. One of the reasons for preferring this arrangement over another – in which the myriapods are the sister-group to the chelicerates – is the possession of a pair of chewing jaws, or mandibles, in all arthropods except the chelicerates. This is interesting because it shows that molecular information has not replaced its morphological counterpart in the business of inferring patterns of evolutionary relatedness. Rather, today we use both types of information. Where they agree, we can usually be quite certain that the tree they suggest is right; where they disagree, the choice of which to believe is not always an easy one.

If the pattern of relatedness among the extant groups of arthropods shown in Figure 14.3 is correct, what does this tell us about the number

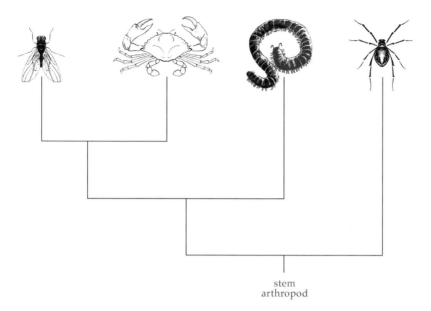

stem
arthropod

FIGURE 14.3 The evolutionary tree of extant arthropod groups that is believed by most present-day biologists. This differs from the tree that was popular for much of the twentieth century, in which myriapods, rather than crustaceans, were the sister-group of the insects. Here, the fly represents insects, the crab crustaceans, the millipede myriapods, and the spider chelicerates.

of invasions of the land by the arthropods? The answer is that it sets a minimum number of such invasions at four: one each by an ancestral myriapod, an ancestral insect, an ancestral arachnid, and an early member of the crustacean group Isopoda ('equal legs') that led to the terrestrial woodlice. However, this number is indeed a *minimum*, and the actual number of arthropod invasions of the land is likely to be higher. There are several species of crab that have evolved to adopt what might be called a quasi-terrestrial existence, for example living on the trees of coastal mangrove forests. Also, many residents of the British Isles will have noticed in their gardens crustaceans that are related to the familiar coastal sand-hoppers. These live under stones and can make spectacular jumps if their home stone is overturned. This appears to be a rather recent invasion of the land by yet another group of crustaceans.

I said earlier that our line of argument would also take us to convergent evolution. It's now time to make that topic explicit. If insects and myriapods invaded the land independently, then their tracheal systems for breathing air must have been evolved convergently. Although this initially seems like an uneconomic hypothesis – one that we should attack with Occam's razor – convergent evolution is actually very common in nature. Evolution really does quite often invent the same thing (well, remarkably similar things, to be more precise) twice or indeed several times.

The above account of arthropod relationships and of the multiple invasions of the land by these animals was based on comparative information on extant forms. But what about fossils, both of those groups that still have extant members and of those (such as the trilobites) that do not? Two geological periods stand out as giving us minimum ages for the origins of the various arthropod groups: the Cambrian (with which we are already familiar; 542–485 MYA) and the Silurian (with which we are not; 443–417 MYA). Interestingly, there are no convincing pre-Cambrian arthropod fossils. Even among the strange disputed creatures of the Ediacaran biota, there are claimed molluscs and annelid worms, but, as yet, no animals that seem likely to be arthropods.

In the Cambrian, when all animal life was aquatic, there are clear examples of fossil crustaceans, along with fossils representing extinct arthropod groups. But it is not until we come forward many millions of years to the late Silurian that we find fossils of terrestrial arthropods. These include the earliest known millipedes, from about 425 MYA, and the earliest arachnids not much later – about 420 MYA. The earliest insects may well also have appeared in the late Silurian, though currently the earliest fossil insect is dated at 410 MYA – about 7 million years after the end of the Silurian period, and thus in the Devonian (see Appendix).

Given that arthropod fossils are usually preserved parts of exoskeleton, perhaps now is a good time to return to that key arthropod feature with which this chapter began. There are many variations in the structure of the exoskeleton among different arthropod groups, but enough similarities to make a brief general account possible, as follows.

The arthropod exoskeleton is composed of non-living material (the cuticle) secreted by underlying living cells, just as the exterior of a typical mammal is composed of non-living material (fur, hair), also secreted by cells that are mostly hidden from view beneath it. In both cases we can call the cells concerned skin cells. However, if we want to be a little more technical, the arthropod cuticle-secreting cells can be collectively referred to as the hypodermis. This is usually just a thin, single-cell layer, but what it secretes is vastly thicker than itself. The cuticle is very complex in terms of its constituents, being composed of a variety of proteins, carbohydrates – such as chitin – and fats, along with molecules that combine two of these – such as large fatty lipoproteins.

What happens to the secreted cuticle in terms of hardening depends both on its position in the body and on the type of arthropod concerned. Some parts of the exoskeleton are hardened by cross-linking of the constituent proteins (a process called sclerotization). This produces hard plates called sclerites. In a centipede, these plates extend along the dorsal and ventral surfaces of the body; however, at the sides, part of the cuticle is unsclerotized and hence much softer (Figure 14.4). In many marine crustaceans, such as crabs and lobsters, the cuticle is hardened further by mineralization – the incorporation of the mineral calcium carbonate into certain layers of the cuticle, making it completely rigid.

Although the exoskeleton is, as noted at the start of this chapter, a key feature in the success of the arthropods, it carries with it a particular disadvantage: it restricts the ways in which these animals can grow. Not

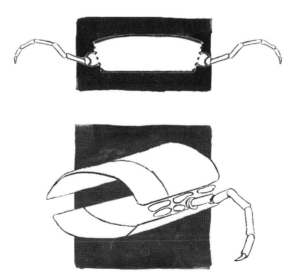

FIGURE 14.4 Two views of a centipede, showing the difference in the arrangement of exoskeletal plates at different points around the circumference of a trunk segment. As can be seen, the dorsal and ventral surfaces have large, integral plates (tergites and sternites respectively), whereas at the sides of the trunk there are little plates of exoskeleton and areas in between them where the cuticle remains more flexible.

only those with the hardest exoskeletons, like crabs, but also those with much softer ones, like spiders, must grow by shedding their old exoskeleton – by the process known as moulting (or ecdysis) – and replacing it with a new, larger one. A typical arthropod grows via a series of such moults, the number depending on the species.

Inconvenient as this may be for the animals themselves, the growth-by-moulting process has been helpful to evolutionary biologists in determining what other groups of animals the arthropods are most closely related to. So far in this chapter, we have only looked at the internal evolutionary relationships of the Arthropoda – how its main subgroups are related to each other. But if we broaden out we can also ask: which animal phyla are the closest relatives of the arthropods?

This question can be addressed in two stages. First, it has long been recognized that two of the minor phyla (in terms of current species numbers) are close arthropod relatives. These are the velvet worms that got a brief mention in Chapter 4 and the water-bears that we will discuss in Chapter 28. The consensus on this issue is such that the term Panarthropoda is often used to include these two groups along with the arthropods themselves. That's fine as far as it goes, but it leads immediately to the second stage of answering the question about arthropod evolutionary relationships: which phyla are closely related to the Panarthropoda?

This is one of the nodes in the animal evolutionary tree regarding which there has been the biggest revolution in thinking about high-level (inter-phylum) relationships in the last two decades. As noted in Chapter 8, before the 1990s there was a consensus that the sister-group of Panarthropoda was Annelida – the segmented worms. The commonality of segmentation between the two groups seemed to be persuasive in this respect. But then along came a paper in the journal *Nature* in 1997 by Anna Marie Aguinaldo and colleagues showing that molecular data supported a different pattern, in which (a) arthropods and annelid worms were only very distantly related; and (b) several other phyla, including Nematoda (the roundworms), were closer cousins of the arthropods. The thing that connected these phyla was growth by moulting. So it seems that segmentation-in-common had been misleading, whereas moulting-in-common was pointing to the true pattern of relationship. This view is now well established – it will be discussed further in Chapter 18 – and the super-group of several phyla of moulting animals now bears the name Ecdysozoa. Since ecdysis is the technical term for moulting, this is an appropriate choice.

The topic covered by this chapter is huge, since arthropods make up such a large fraction of the animal kingdom. Given how many aspects of arthropod biology there are, we have discussed only a small selection of them. Evolutionary relationships, exoskeletons, breathing systems, life-cycles, larvae and invasions of the land are all fascinating topics, to be sure. But are not feeding ecology, circulatory systems and the morphology of the head (among many other topics) equally interesting? The short answer to this question is 'yes', with the qualification that what any one person finds most interesting about a group of animals will probably not be the same as what another person finds most interesting. However, for those who would like to take a more comprehensive guided tour of the arthropods there are many good books from which to choose, including the one published in 2013 entitled *Arthropod Biology and Evolution*, edited by Italian zoologist Alessandro Minelli and his colleagues.

15 Extinction

Individual animals are born and die; the same applies to species. Admittedly, the birth of a species through speciation – which we looked at in Chapter 5 – and its death through extinction – our subject here – are very different processes from individual births and deaths. Nevertheless, species, like individuals, are bounded in time: they have a start-point, a finite duration, and an end-point. The duration of a species' existence would be impossible to predict at its inception: it might be less than a million years, or more than 10 million. The fossil record suggests that a few species remain in existence for over 100 million years – these are sometimes called living fossils, because the extant animal of today resembles so closely a fossil form that may be its ancestor of the distant past. The arthropod called the horseshoe crab is a case in point (Figure 15.1).

Extinction is a multi-level phenomenon. In the above introduction, I focused on the level of the species. But we can use the word extinction to refer to groups of animals both above and below the species level. This chapter will include both of these usages. And in fact we'll start with groups below the species, because this helps to reveal that ultimately extinction is an ecological process, just as much as it is an evolutionary one. In a sense, the mechanism of extinction is ecological while its result is evolutionary.

Any species, at any moment in time, consists of a mosaic of geographically separate, or quasi-separate, populations. The overall extent of these populations is referred to as the geographical range of the species. Ranges vary enormously in size. For example, each bird species belonging to the group called Darwin's finches has a geographical range that extends only part-way through the Galapagos archipelago of islands and islets. And each species of lemur is restricted to part of the island of Madagascar. In contrast, the red fox is an example of an animal with a holarctic range, which means that it occupies all the northern continents – North America, Europe and most of Asia. The human species occupies every continent on Earth. Omnipresent species like our own are described as having cosmopolitan distributions; those that are restricted to a single

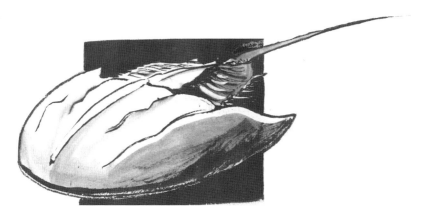

FIGURE 15.1 A living horseshoe crab. These are marine arthropods, superficial inspection of which might lead an observer to think of them as being some kind of crustacean. However, detailed observation reveals that they are more closely related to spiders and scorpions, and so belong to the arthropod group Chelicerata. They are described as living fossils because they closely resemble fossil forms from more than 400 MYA. However, this term should be interpreted with caution: it is unlikely that the earliest and extant forms belong to the same species.

place are described as being endemic to that place – often an island in the case of terrestrial species or a lake in the case of freshwater ones.

When the number of animals in a particular geographical population of a species dwindles to zero, ecologists call this a *local* extinction. Many such extinctions have occurred of red squirrel populations over much of Britain and Ireland in the last century, due, apparently, to a virus spread by the grey squirrel that was introduced from the United States and has spread rapidly. The grey seems to be unaffected by the virus that it carries, which is lethal to the reds. But the red squirrel is not yet extinct on either island; and even if it becomes so it will not be extinct as a species, as long as mainland European and Asian populations remain. (They might not: immigrant greys in northern Italy are already displacing reds there too.)

The degree to which a local extinction represents a step along the road to global extinction – the ultimate end of a species – depends, of course, on the extent of its geographical range. For an endemic species restricted to a single small island, a local extinction caused by some very localized ecological event – such as a forest fire – will constitute a global extinction too. But cosmopolitan species are much harder to send into oblivion. And there's a sliding scale from one extreme to the other.

Let's now move up and examine a level above that of the species. When palaeontologists examine the pattern of extinctions, one of the levels that they use is the family. Some family names are familiar – for example Felidae (the cat family) and Canidae (the dog family) – while others are not. Anyhow, just as a species only becomes extinct when all its constituent populations have disappeared, so too a family only becomes extinct when all the species of which it was made up have gone.

These two situations are not quite as similar as I have made them sound, in part because while a species can be defined objectively, in terms of interbreeding, a family cannot. Families – and other levels of group above the species – are subjective constructs. Although this fact has led to some calls for abandoning them altogether, that would lead to all manner of problems: better subjectivity than chaos, at least in my view.

Another reason why the population–species relationship is not the same as the species–family one is that while the different populations that make up a species are all, by definition, in different places, the different species that make up a family may not be. For example, one of the main families of lemurs is called (appropriately) the Lemuridae. This consists of about 20 species, all endemic to Madagascar. In other cases, the species that make up a family are much more widely scattered. For example, the gibbon family Hylobatidae also consists of about 20 species, but these extend across much of the mainland of south-east Asia, as well as across parts of the islands of Borneo, Sumatra and Java. The likelihood of extinction of a family thus depends on at least two things: its number of constituent species and the degree to which they are geographically spread.

During any span of geological time, species originate by speciation and disappear through extinction. Because families are groups of related species, they too come and go when long enough periods of time are considered. Extinction is thus a normal part of the overall evolutionary process through which the array of species and families populating the Earth changes. And while in a sense extinction is a negative thing, it is also (less obviously) a positive one, because the extinction of a species with a particular ecological role may result in the evolution of a new species. As the saying goes, nature abhors a vacuum.

When we look at the rate of extinction, we see that it varies considerably. This is hardly surprising: most (arguably all) ecological and evolutionary processes exhibit rate variation. But exactly how we should interpret the pattern of rate variation is problematic. The percentage of species or families going extinct every 5 million years looks like the profile of a range of hills and mountains when we consider the long span of geological time from the start of the Cambrian period (542 MYA) to the present. This span of time is long enough to allow more than 100 such estimates, or indeed more than 500 estimates for the per-million-years rate, to be made.

The highest mountains are often referred to as mass extinction events. But how high does a mountain have to be in order to be thought of in this way? According to some sources, it doesn't make sense to single out any mountains – they are just extreme blips on a generally messy pattern of change. Other sources admit that at least two mass extinctions should be recognized, while others claim that there have been six or more. And, regardless of these differing opinions of *past* extinction, there are also divergent views on the extent to which we humans are causing a mass extinction event that is just starting.

I'm going to take the view that at least two peaks in extinction rate are worthy of being described as mass extinction events. These were the peaks that occurred at the ends of the geological eras known as the Palaeozoic (old animals) and the Mesozoic (middle animals). I have dealt more with geological periods, especially the Cambrian, than I have with eras, so a brief word of explanation is in order here about the relationship between the two.

From the start of the Cambrian to the present there have been 12 periods, but only three eras. The correspondence is as follows: six periods in the Palaeozoic, three in the Mesozoic and three in the era of so-called recent animals, the Cenozoic (pronounced 'seen-oh-zo-ik'). These three eras and 12 periods are shown in a diagram in the Appendix; but right here I'm following my policy of keeping the main text as jargon-free as possible.

It's hardly surprising that the Palaeozoic and Mesozoic eras ended with mass extinction events: after all, these events were used to define the eras in the first place. But what is surprising is the magnitude of the extinctions. We will measure their magnitudes at two levels: that of the species and that of the family, thus matching the above introductory remarks. We'll also label each with a well-known higher-level group that disappeared in the event concerned.

The end-Palaeozoic event (about 250 MYA) is thought to have resulted in the extinction of more than 90% of species and more than 50% of families. It was also the event in which the trilobites finally disappeared, after making it through several earlier peaks in extinction rate. The end-Mesozoic event (about 65 MYA; sometimes called the K/T event, for reasons that can be found in the Appendix) was less intense, even though it is more famous, given that it included the extinction of the dinosaurs. In this event, about 70% of species and 20% of families are thought to have become extinct.

The causes of mass extinctions have been fiercely debated, and none more so than the dinosaur-killing end-of-the-Mesozoic event. The hypothesis that seems to be in the ascendancy is that this event was caused by the impact of a large extraterrestrial body, probably an asteroid, with the Earth, approximately 65 MYA. There is considerable evidence in support of such an impact. First, over many parts of the Earth, the chemical element iridium is found at a depth within the rocks that indicates an age of about 65 million years. Iridium is a scarce element in the crust of the Earth in general (even rarer than platinum), but it is much more common in meteorites. Analysis of these has shown concentrations of iridium that are more than 100 times greater than in terrestrial rocks. So if an asteroid or other large extraterrestrial body impacted Earth 65 MYA, its disintegration in the collision would have scattered iridium widely, and could easily have produced a layer of the sort that we observe.

The second piece of evidence for impact is the crater itself. One of the largest craters on Earth, the 160-kilometre wide Chicxulub crater, partly buried under the Yucatan peninsula of Mexico, dates from about 65 MYA (Figure 15.2). Of course, the fact that this is approximately the same age as the worldwide iridium layer could be a coincidence of some sort; but that doesn't seem likely. So the hypothesis that an asteroid or other large object collided with the Earth, disintegrated, and scattered iridium and other material around the globe, has much to support it.

What happened next is much more uncertain, though. The key question here is this: what is the link between the impact and the mass extinction event? Scenarios that have been proposed include an extended period in which there was so much smoke and debris in the atmosphere that sunlight was severely reduced or eliminated, making

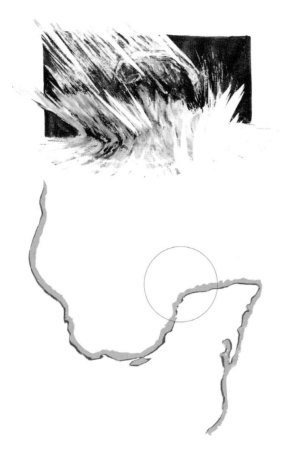

FIGURE 15.2 Artist's impression of the collision of an asteroid with the Earth approximately 65 MYA (top). Such a collision probably caused the large Chicxulub crater that now exists in Central America. Part of the crater is on the Yucatan peninsula of Mexico, while part of it lies under the sea (bottom).

photosynthesis by plants impossible. This would in turn result in a lack of food for herbivores, and thus also a lack of meat for carnivores. Detritivores would have a field-day. But how long was this extended period? And how did non-dinosaurian herbivores and carnivores survive? It has been suggested that because the mammals of the day were small creatures they might have been able to find enough food. But what about the crocodiles? Why did these survive this great mass extinction event? Crocodiles are smaller than the large dinosaurs but larger than the small ones. Was their survival in part due to their ability to live for extended periods under water, as some palaeontologists have argued? And why did diverse groups of tiny marine invertebrates go extinct? There are still many questions to be answered.

Now this might seem like a strange statement, but in one sense the dinosaurs are not extinct. Recall that a family can only be said to be extinct when all of its member-species have perished. This is true of higher levels of taxon too, all the way up to the phylum. Let's now explicitly link these levels of taxon with evolutionary trees. We began this process in Chapter 7 (entitled *How to make a tree*); and we noted there that the simplest type of tree, giving just the temporal sequence of lineage-splitting events and no other details, is called a cladogram. Connected with this term is another one: *clade*. In the context of a cladogram, or indeed of evolutionary trees in general, any stem lineage and all its descendant lineages are collectively referred to as a clade. Therefore, each family is a clade; so too is each phylum. This is simply another way of saying that valid groups of animals must be, as it was put in Chapter 7, *monophyletic* ones.

The problem with Dinosauria (when interpreted as the group containing only what most people would call dinosaurs) is that it is not a complete clade; rather, it is a clade with a bit missing – a paraphyletic group. This can be seen in Figure 15.3, where it's apparent that, in order for there to be no bit missing from the clade, birds need to be included in it. They're not even one of the most basally branching groups within the clade; so they are more closely related to some dinosaurs than those dinosaurs are to some other dinosaurs. Thus there really is no way to exclude birds from being part of this clade, for which the name Dinosauria then seems inappropriate – unless you regard birds as dinosaurs, which some biologists now do; they then refer to the creatures that most folk call dinosaurs as the 'non-avian dinosaurs'.

This is one kind of problem that we need to confront when thinking about extinction. But there's another one too, this time rather more practical than conceptual. How do we know that a group (be it a species, a family, or some other level of taxon) really is extinct, rather than being still extant but so rare and/or occupying such difficult-to-access habitats that we can't find it? This problem is clearly minimal with the (non-avian) dinosaurs. The probability of a *Tyrannosaurus rex* being discovered in a remote part of South America – or elsewhere – can be regarded as vanishingly low. But in some other groups of animals, the problem is very real.

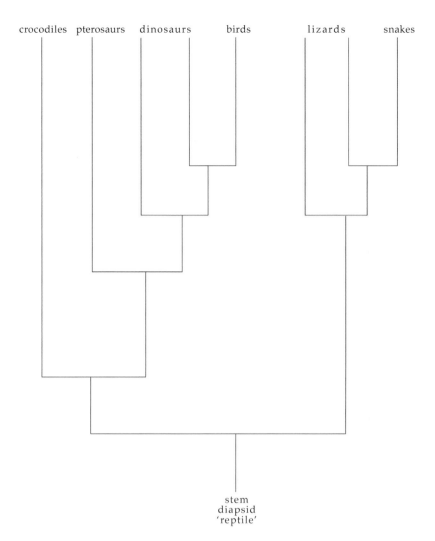

crocodiles pterosaurs dinosaurs   birds   lizards   snakes

stem
diapsid
'reptile'

FIGURE 15.3 A cladogram showing the pattern of relationship among different groups of reptiles. Note that birds are more closely related to dinosaurs than some of the latter are to each other. Thus a group called Dinosauria is a paraphyletic one unless it includes the birds. Likewise, it can be seen that, without the snakes, lizards constitute a paraphyletic group.

The best example of this is the fish called the coelacanth (Figure 15.4). The story of the discovery of living specimens of this supposedly extinct fish has been told in detail by zoologist Keith Thomson in his 1991 book *Living Fossil* – near the start of which he has very sensibly devoted an entire page to a single question and its answer: how do you pronounce it?

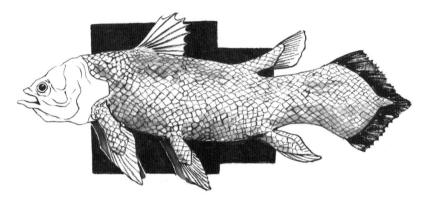

FIGURE 15.4 The fish called a coelacanth that was thought to have been extinct for at least 65 million years before one turned up in the catch of a fisherman in South Africa in the 1930s. More than 200 additional specimens have since been discovered. Note the fleshy paired pectoral and pelvic fins; we'll discuss these further in Chapter 21.

"seel-uh-kanth". I'll just give a very brief account here; those who are sufficiently interested to pursue the story further can read Thomson's book.

Coelacanths were thought by late-nineteenth- and early-twentieth-century palaeontologists to have been creatures of the Palaeozoic and Mesozoic eras, which finally died out, along with the dinosaurs, in the end-Mesozoic extinction event of 65 MYA. The most recent fossils known were from several million years before that date. The fossils had been reasonably well described by 1930. There were many of them; the discovery of another fossil would not have surprised anyone. But what happened in December 1938 was much more dramatic than the finding of a new fossil coelacanth: the finding of an extant – albeit dead – one.

This coelacanth specimen was caught by a fishing boat off the coast of South Africa. The curator of the East London Natural History Museum in Cape Province, Marjorie Courtenay-Latimer, regularly visited the nearby fishing port to inspect the catch. Often, nothing of much interest was to be found. But on a particular occasion in 1938, among the pile of sharks, dogfish and miscellaneous other sea creatures that were left over after the saleable fish had been taken away, there was something quite remarkable: a strange five-foot long fish of a bluish colour with iridescent silver markings.

Courtenay-Latimer suspected that this specimen was something out of the ordinary – but what? She consulted James Smith, a leading fish specialist. To cut a long story short, their interaction resulted in a paper being published by Smith in the journal *Nature* in March 1939 entitled "A living fish of Mesozoic type". A second specimen was caught in December 1952 – fourteen years after the catch of the first one. Since then, about 200 more specimens have been found. A fish that was thought (you might even say 'known') to be extinct was alive and well, albeit perhaps not with very high population numbers – Thomson describes it as being "probably endangered".

Now let's turn our attention to another positive story about extinction – this time not about a perception of extinction that turned out to be wrong but rather about the positive effects that real extinctions can have on the animals that remain. So we're back to that point about nature abhorring a vacuum.

A more scientific way to make this point is that most species, for most of the time, are engaged in what Darwin called a "struggle for existence" (this being the title of chapter 3 in *On the Origin of Species*). But after an extinction, the ecological role, or niche, of the species that has become extinct is, in a sense, empty. The resources that it utilized are there for the taking. Any remaining species that can evolve so that it is able to utilize those resources is likely to be favoured by natural selection. And think of the situation after a mass extinction event when huge numbers of species have perished. The opportunities for rapid evolution of the survivors are many and varied. It has often been said that if it were not for the extinction of the dinosaurs, the mammals would never have undergone the great evolutionary radiation that they did, and that we humans might never have come to be. Of course, such hypotheses can't be tested because we don't have a parallel planet that's the same in all respects except for the dinosaurs surviving the end-Mesozoic extinction event. But at least it's a *plausible* hypothesis; and that's a sobering thought.

16 Mouth first, mouth second

The shape of the animal kingdom, as we see it today, has been sculpted both by origins and by extinctions. Having examined extinctions, we now return to the more positive side of the evolutionary process: origins. But what, in an evolutionary context, is an origin? Every species of animal that has ever lived began by evolutionary modification of its parent species. And all animal groups (or clades, as we now think of them) have their origins within other clades – such as the birds within the dinosaur clade, as we noted in the previous chapter. There is no such thing as the magical *de novo* origin of any new animal species, or any new animal group, devoid of a history of ancestors.

Having said that, some origins are inherently more interesting than others, and for at least two reasons. At the level of the individual animal species, there is more interest in the origin of *Homo sapiens* than of any other species for the obvious reason that we are particularly fascinated by our own roots. And this kind of self-interest can extend from individual species to higher-level groups – for example, going back through time, to the origin of the primates, the mammals and the vertebrates.

Another reason for taking an interest in the origin of a clade is connected with its size. Large clades, containing a sizeable chunk of the animal kingdom, are especially interesting. While the vertebrates provide an example of a clade that is interesting for both reasons – self-interest and size – other large clades, for example the arthropods, are of interest for size reasons alone, since we do not belong there.

Clades that are large tend also to be old. To get to grips with the issue of a clade's age, we need to grapple with different spans of time. It has long been conventional to separate ecological time (up to a few hundred years) from evolutionary, or geological, time (usually measured in millions of years). However, some other form of terminology is needed to separate the origins of the major clades of the animal kingdom – phyla and subphyla – from the origins of individual species. It seems that all the animal phyla, including chordates, arthropods and molluscs, arose in the Ediacaran or Cambrian periods, together spanning from about 635 to 485 million years ago (MYA). Increasingly, such ancient evolutionary origins are referred to as happening in deep time – even though there is no official criterion for what is deep and what is not. Clearly, the earlier an evolutionary origin, the deeper it is – there is a continuum rather than a dichotomy in this respect. Nevertheless, to get what's often called a ball-park view of how the phrase deep time is used: it tends to be restricted to events that took place before the end of the Palaeozoic era, approximately 250 MYA. If we want to adopt an arbitrary definition of deep time, that one's as good as any.

To summarize: deeply originating clades with many constituent sub-clades and species are, other things being equal, the ones that evolutionary biologists find most interesting. An example should help to illustrate this point. In Chapter 6, we looked at the group containing the jellyfish and their allies (the phylum Cnidaria) and noted that it diverged from the lineage that led, via the urbilaterian (Chapter 8), to most of the animal kingdom. The cnidarian clade deserved examination because, among other things, it had a deep origin and has undergone an extensive radiation – into not just many kinds of jellyfish but also sea anemones, corals and several other forms. Together, there are about 10,000 species. I have not, however, devoted a chapter (or indeed any prior mention) to another deeply branching animal clade – the 'flat animals' or, more officially, the Placozoa. Although recognized as a phylum, this group contains just a single species, at least on the basis of current knowledge. The animals of this species are flat, as their name suggests (just two cell-layers thick) and small – they are irregular in shape, with the longest axis being typically about 1 millimetre. A common habitat is the inter-tidal zone of a tropical ocean. You'll probably never see one in the flesh, so a placozoan is pictured in Figure 16.1.

Although assigning importance on the basis of clade size is some-what subjective, it's a pragmatic way of dealing with the problem of trying to understand the structure of the animal kingdom without drowning in detail. So let's pursue this approach further and look at a key deep split in the Bilateria: the split that produced the two very large clades that we call protostomes and deuterostomes. These two words translate, respectively, as mouth first and mouth second – hence this chapter's title.

Here, we are entering the territory of comparative embryology – which we'll explore more widely in the following chapter. 'Mouth first' refers to an embryo making the mouth-end of its digestive tract before

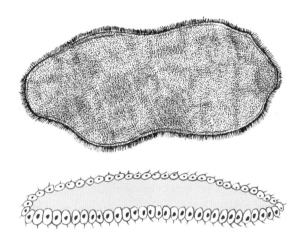

FIGURE 16.1 One of the simplest of all animals: a placozoan, seen from above (top) and in cross-section (bottom). These tiny creatures are about a millimetre long and their bodies are just two cell-layers thick.

the other end (the anus). 'Mouth second' refers to the opposite embryonic sequence – making the anus-end first and the mouth later.

The terms *protostome* and *deuterostome* were first introduced by the Austrian zoologist Karl Grobben in the early twentieth century. Grobben essentially divided the main part of the animal kingdom (bilaterian animals) into two, by using the criterion of whether the mouth or the anus was made first in embryogenesis. As it turns out, there is a third possibility, that the two are made more-or-less simultaneously, which complicates the issue. However, we'll get to that later.

For now, a reminder is necessary of one of the key early processes in the embryonic development of most animals, which we looked at in Chapter 13, namely gastrulation. To describe one of the main things happening during this process, I used the analogy of poking a finger into a balloon, thus producing the beginnings of an internal tube. Although poking too far into a real balloon would result in it bursting, poking a long way into an embryo can result, eventually, in the forming internal tube arriving at the other end of the embryo (which elongates during this process, so that it comes to have an antero-posterior axis). In protostomes, Grobben thought that the initial poking in makes the mouth, while the final breaking through at the other end makes the anus. In deuterostomes (including us), it's the other way round (Figure 16.2). So we could be called anus first, but mouth second is perhaps a little more dignified.

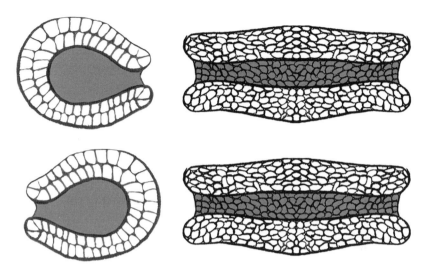

FIGURE 16.2 Two types of gastrulation: mouth-first (top, anterior to the right in both illustrations) and mouth-second (bottom, anterior again to the right). In both cases, the connection between the diagram on the left and its (later) counterpart on the right is elongation of what will become the anterior–posterior axis and formation of a second opening at the opposite end of the embryo from the first one. For complications, see text.

The bilaterians make up about 99% of the animal kingdom, and the split between these two groups of bilaterians – the protostomes and deuterostomes – is the deepest major split within the Bilateria, as far as is currently known. Yet these names are unfamiliar, not only to the general public, but also to many biologists. Why?

The answer to this question probably lies in the fact that no obvious features of the adults unite either group. This is in contrast to, for example, vertebrates and arthropods, with their familiar endo- and exo-skeletons. There is no general protostome body plan, nor a deuterostome equivalent. So we have no mental image of a typical member of either group – because there is no such creature.

This is a good point, then, at which to give some examples of animal groups that we can mentally picture, and to indicate which belong to the mouth-first group and which to the mouth-second. I've already said that we humans make our anuses before our mouths during embryogenesis, so we are clearly deuterostomes. And that goes for all of the vertebrates and indeed all of the chordates. So our own phylum is one of those that belong to what can be called the super-group Deuterostomia. In fact, ours is the biggest phylum, in terms of species numbers, within this group. The second biggest is the Echinodermata – starfish and their allies. Taking the same approach to the protostomes, the biggest phylum in that super-group is the Arthropoda, with the Mollusca being second.

Notice the very great differences among these four phyla. If you were asked to group them into two pairs on the basis of the forms of the adult animals, it would be hard to know how to proceed. If we use C for chordates, E for echinoderms, A for arthropods and M for molluscs, then the three possible pairings are as follows, starting with the right one: 1, ((C + E), (A +M)); 2, ((C + A), (E + M)); 3, ((C + M), (A + E)). Without embryological or molecular evidence, none of these looks any more plausible than the others. Indeed, given that extant echinoderms have such a unique body plan based on five-fold symmetry, it would seem more sensible to propose a pattern of interrelationships that does not take the form of two pairs, but rather has the echinoderms as an out-group to the other three: ((C + A + M), (E)).

When we look beyond these four phyla and ask how the rest are split between the mouth-first and mouth-second groups, we find a great asymmetry. Of the 30 or so bilaterian phyla, most of them are protostomes, while only a few are deuterostomes. All of the other bilaterian phyla we've examined so far in this book (e.g. the segmented worms, the penis-worms, and the roundworms) are protostomes; however, the deuterostomes are not wormless – the phylum Hemichordata, including the acorn worms, belong with the echinoderms and ourselves in the Deuterostomia.

So far, if you haven't previously encountered this deep split in the Bilateria and the names of the groups that accompany it, you have only had my word to take for its reality. And I wouldn't blame anyone who is at this stage beginning to wonder if I'm talking nonsense. Grouping animals in the way indicated above seems to make little sense from the perspective of the forms of the adult animals, as already noted. So the embryological and molecular evidence must be very compelling if we are to believe it. Luckily, it is – well, some of it anyway.

The evidence from comparative embryology was available in the early years of the twentieth century; and it was on the basis of this that Karl Grobben based his original split. The molecular evidence did not become available until almost a century later. We'll take these two types of evidence in the order in which they became available.

Embryologically, protostomes and deuterostomes are often said to differ in important ways during *several* stages of early development. These stages include gastrulation, of course, since this is the stage at which the digestive tube pushes its way through the embryo and thus the stage at which a mouth-first or mouth-second arrangement can become apparent. But they also include the stages immediately before and after gastrulation, namely cleavage and neurulation.

Before we look at those other stages, it's worth pausing for a moment here to reflect again on the claimed primary difference related to gastrulation. There are several problems in claiming that all the so-called deuterostomes do actually make their anus first and all so-called protostomes their mouth first. And they all have to do with what is called taxon sampling. When you are dealing with a very large group of animals, a claim that they all do something in the same way – for example making their anus before their mouth in embryogenesis – needs to have a sound basis. This should include evidence from all the major subgroups of the group concerned.

Here is one example of this problem, in relation to the consistency of embryonic pattern among the different subgroups within the Deutero-stomia. The claim that a vertebrate embryo makes its anus first and its mouth second should not raise any eyebrows; but a similar claim made about their deuterostome cousins, the echinoderms, should. Since a typical echinoderm, say a starfish, has a blind-ending gut with only a single opening to the outside, which functions as both mouth and anus, a claim about which of these is made first seems bizarre. There is, however, a solution to this apparent puzzle. In echinoderms, embryonic processes, gastrulation included, lead to the production of a larva. Most echinoderm larvae have a through-gut with a mouth and anus; and indeed they are bilaterally symmetrical. So being mouth-second is not illogical after all. (The echinoderm adult is constructed, during metamorphosis, from a small piece of tissue called the *rudiment*, usu-ally located on the left-hand side of the larva; this developmental process has been much studied in sea urchins.)

In fact, problems about claimed general patterns of gastrulation related to taxon sampling are more acute in the protostomes than in the deuterostomes. With the above proviso about larval versus adult echinoderms, a general statement that all deuterostomes make their anus first and their mouth second in embryogenesis is not too problematic. However, a statement that all protostomes do the opposite is problematic in the extreme. This is because some animals belonging to the protostome group actually have a mouth-second developmental sequence, while others form the two openings more-or-less simultaneously by a process called amphistomy. For details of these and other complexities in the 'mouth first, mouth second' story, see the chapter by Andreas Hejnol and Mark Martindale in the very stimulating multi-authored 2009 book *Animal Evolution: Genomes, Fossils and Trees*, edited by Max Telford and Tim Littlewood.

Problems of taxon sampling can also be found in relation to claims that are sometimes made about differences between protostomes and deuterostomes in cleavage and neurulation. In relation to cleavage, it is sometimes said that deuterostomes exhibit radial cleavage (the type that was pictured in Figure 13.2), while protostomes exhibit a variant type called spiral cleavage, in which, at the eight-cell stage, the four smaller cells do not lie directly above the four large ones, but rather are displaced, in either a clockwise or an anticlockwise way, relative to those large cells. However, this generalization does not hold up to scrutiny. Many protostomes do indeed have spiral cleavage, but many do not. In fact, the biggest single group of protostomes in terms of species numbers, the insects, exhibit a type of cleavage that is neither radial nor spiral.

We now turn from the developmental process that precedes gastrulation to the one that follows it. In the development of vertebrates, the formation of the rudimentary central nervous system, which will eventually become the brain and spinal cord, is a sufficiently prominent process for it to be named separately from the formation of other organs, such as the heart and liver: as we noted earlier, it's called neurulation. It takes the form of a thickening that is externally visible along the dorsal midline (Figure 16.3). There is no major movement in the position of the spinal cord as it develops further, so in adult vertebrates it, like its embryonic forerunner, is a dorsal midline structure.

If we look at major protostome phyla, such as arthropods and molluscs, we typically find that both in embryos and in adults the main antero-posterior nerve cord runs along the *ventral* midline. This fascinating difference was noted by the French comparative anatomist Etienne Geoffroy Saint-Hilaire in 1822, leading him to propose (and to be ridiculed for) the idea that a vertebrate is a sort of upside-down

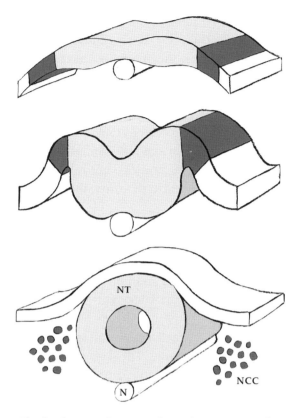

FIGURE 16.3 The developmental process of neurulation in a vertebrate embryo. The neural plate (top centre in top picture) thickens and buckles, becoming U-shaped in cross-section. Then the ends of the U fuse to produce the neural tube (NT), the forerunner of the central nervous system, which lies just above the notochord (N). As these changes are happening, a group of cells, the neural crest cells (NCC), migrate away; they will end up contributing to many structures, including non-neural ones.

arthropod. However, at that time the super-groups protostomes and deuterostomes had not yet been recognized or named, so Geoffroy's comparison could not be seen in that light. Now that it can be, we can ask the question: is a ventral nerve cord common to all protostomes and a dorsal one to all deuterostomes?

The answer to this question is a qualified yes. We can't seriously expect the pattern to apply *universally*, as evolutionary patterns are rarely if ever so neat and tidy. But the pattern does indeed apply *generally*. Most protostomes have a ventral nerve cord; and most deuterostomes have a dorsal nerve cord. However, in relation to the latter generalization, it must be admitted that some deuterostomes have no central nerve cord at all. This is the case in echinoderms because, having lost their bilateral symmetry, they no longer have a brain or a spinal cord, just a diffuse nerve net.

This difference in which side of the animal its main nerve cord is on – the back or the front – raises an interesting evolutionary question. Assuming that protostomes and deuterostomes had a common ancestor in deep time – and no biologist seriously doubts this – what was the layout of its nervous system, and how did later animals in one or both of the main lineages leading forward in time from that ancestor change to a different layout?

There are two possibilities. The first is that the last common ancestor (LCA) of protostomes and deuterostomes had a diffuse nerve network, with the two descendant lineages independently evolving a central nerve cord, and doing so, either for adaptive reasons or by chance, on different sides of the body. This seems unlikely because the urbilaterian (Chapter 8) probably already had a central nerve cord, and either it or one of its descendants was presumably the LCA of the protostomes and the deuterostomes. The second is that the urbilaterian had either a dorsal or a ventral nerve cord, in which case either the ur-protostome or the ur-deuterostome evolutionarily inverted its dorso-ventral layout. That seems a more plausible hypothesis, but nevertheless a vague one because the exact nature of any such inversion is not clear.

So, to summarize the embryological evidence, protostomes and deuterostomes differ at the earliest stages of embryogenesis – cleavage, gastrulation and neurulation – but the differences between them are not without complications. Because of increased taxon sampling of the two groups, many more complications are known today than when Karl Grobben proposed the two super-groups over 100 years ago. If all we had now was the embryological evidence, we might have begun to doubt the wisdom of dividing the bilaterian animals into protostomes and deuterostomes. However, the molecular genetic evidence that has accumulated over the last three decades has confirmed the soundness of these super-groups, despite the embryological complications. I should perhaps admit that a few minor phyla have shifted position, compared to Grobben's placing of them, on the basis of the molecular evidence, but all the major ones have stayed put.

Given that a typical animal genome has about 20,000 genes (with a range of variation of at least plus/minus 5000 around that figure), there is a very large number of genes to choose from to use as characters to discern the pattern of evolutionary relatedness – arguably even greater than the number of embryological and adult morphological characters. So how do we decide which gene(s) to use?

The answer is two-fold. To use a particular gene to look at patterns of divergence in deep time, we need to choose a gene that evolves rather slowly. But we should choose more than one such gene, because any evolutionary tree based on a single character constitutes a very risky hypothesis of a pattern of relatedness. Studies of slow-evolving genes have consistently produced the groupings of phyla that we recognize as the protostomes and deuterostomes. Taken together with the earlier embryological evidence, albeit this has changed a bit due to improved taxon sampling, the results of these studies confirm beyond reasonable doubt that the split between protostomes and deuterostomes was one of the deepest major divergences within the animal kingdom, and probably the deepest of all within the Bilateria.

17 Comparing embryos

When first examining animal development in Chapter 13, we noted that a pioneer of the subject of embryology was the Italian Hieronymus Fabricius, whose studies on the development of the chick were published posthumously in 1621. This work can best be considered as *descriptive* embryology – it involved a focus on a particular animal and gave as detailed a description of its embryogenesis as the techniques of the time would allow.

It was not until about two centuries later that embryology became truly *comparative*. This aspect of embryology is clearly related to evolution, as was recognized by Darwin, who, in 1859, devoted part of chapter 13 in *On the Origin of Species* to their relationship. An often-quoted statement that Darwin makes there is: "community in embryonic structure reveals community of descent." We might do well to examine this statement carefully, to try to come up with a modern version of it, and to see what course of logic such an attempt will set in train.

What Darwin was saying is that if two kinds of animal have similar patterns of embryogenesis this shows that they have a common ancestor. In other words, he was using comparative embryology as a source of evidence for the reality of evolution. That was of course a very sensible thing to do back in the 1850s; and it was part of Darwin's broader strategy of drawing on all possible sources of evidence in favour of evolution and against the alternative of the immutability of species. But the fact of evolution is no longer in question, so the way we should look at Darwin's statement today is very different from the way in which his contemporaries looked at it.

We now acknowledge that there is community of descent throughout the living world. This goes hand-in-hand with the idea of a *last universal common ancestor* (or LUCA) that was first mentioned in Chapter 2. But this does not mean that there is "community in embryonic structure" wherever we look for it. There is not much similarity between the embryos of sponges and crocodiles, yet they do indeed share a common ancestor. So an important point to note is that there can be community of descent with or without embryonic similarity.

This point brings us to the issue of closeness of evolutionary relationship. Generally speaking, we find most embryonic similarity when the forms we are comparing are closely related. Thus, for example, although there are a few similarities between the embryos of humans and starfish, as we saw in the previous chapter, there are many more similarities between the embryos of humans and cows.

So, rather than having two pairs of possibilities – presence or absence of embryonic similarity and presence or absence of evolutionary relatedness – we have two continua. These are the *degree* of embryonic similarity and the *closeness* of evolutionary relationship. In general, we should expect a correlation between these two. In other words, the closer, in an evolutionary sense, are the two animals we are comparing, the more embryonic similarity we should expect to find. In general this turns out to be true, though there are, as ever in the living world, plenty of 'exceptions that prove the rule'. Also, there is an important distinction between *embryonic* and *developmental*. The former refers to events occurring within the protection of an egg-casing of some kind or within some other protective casing such as the mammalian womb. The latter is a more general term; it includes larvae, where developmental events are taking place against the background of being exposed to varying environmental conditions.

It is generally the case that larval development is less clearly correlated with closeness of evolutionary relationship than is embryonic development. This can be seen by comparing the embryos of the very closely related species of sea urchins studied by the American biologist Rudolf Raff and his colleagues. Sea urchins have planktonic larvae that are either feeding or non-feeding. The former have feeding arms, while the latter – which are nourished by an internal yolk supply – do not. These latter larvae are rather featureless in comparison (Figure 17.1). The two embryos pictured belong to sister-species that separated from each other only recently in evolutionary terms; yet clearly they are very different from each other. Moreover, non-feeding larvae seem to have evolved from feeding ones several times in sea urchins – another example to add to our already long list of cases of convergent evolution. This means that, if we were to randomly choose four species for comparison, two with feeding larvae and two with non-feeding ones, we could easily find, just by chance, that larval similarity was associated with *distance*, rather than *closeness*, of evolutionary relationship – the opposite of what we would have expected.

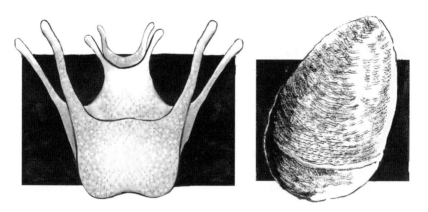

FIGURE 17.1 Feeding (left) and non-feeding (right) sea-urchin larvae. Note the completely different larval forms, with the feeding larva having prominent appendages known as feeding arms and the other larva, which is nourished by a yolk supply, lacking these.

This cautionary tale serves as a reminder that it is important to avoid sweeping statements such as 'closeness of relationship always goes hand-in-hand with developmental similarity.' And Darwin, cautious as ever, was careful not to make an over-general statement of this kind. His use of "embryonic" in his statement that developmental similarity and common descent go together was wise. And the restriction this entails was almost certainly deliberate, because Darwin goes on (in his chapter 13) to point out that, in animals with larvae (and hence indirect development and a complex life-cycle), "the adaptation of the larva to its conditions of life is just as perfect and as beautiful as in the adult animal." He continues: "From such special adaptations, the similarity of the larvae ... of allied animals is sometimes much obscured."

To summarize, then: in animals with direct development and no larval stage, such as those of the mammal/reptile/bird clade, we see a clear link between closeness of evolutionary relationship and similarity in embryogenesis. The link is not universal or exact; but it is general and, from a statistical point of view, strong. In contrast, in animals with a dispersive larval stage, such as sea urchins, the link between closeness of evolutionary relationship and early developmental similarity is much messier, and, depending on exactly which species are chosen for comparison, may, to use Darwin's word, be more or less obscured.

In the above logical progression of thought, we moved from two dichotomies to two continua – those of degree of embryonic similarity and closeness of evolutionary relationship. But now we need to add a third continuum: developmental time. It is perhaps useful at this point to stress that in all of the above discussion of *embryonic* similarity no mention was made of which embryonic stage was being compared between two (or more) animals. Indeed, the lack of reference to particular stages suggests that the focus of the argument was on some sort of *overall* degree of similarity between embryos.

But this is not enough. All embryos have a certain duration in time, bounded, in most cases, by the starting point of fertilization and the end-point of hatching or birth. Taking any two kinds of animal for comparison, the degree of embryonic similarity may change considerably through that span of time. Also, interacting with this complexity, there's another one: embryonic duration may itself change with evolution. Thus in a comparison of, let's say, embryonic stage 1 between two species of animal, we need to decide whether the definition of this stage should be in units of absolute time or in units defined by developmental processes such as cleavage, gastrulation and so on.

Making the assumption that these problems can be overcome, how can we refine our view of the relationship between evolutionary closeness and embryonic similarity, by adding in the third continuum of developmental time? Well, we can for a start catch up with the great comparative embryologist Karl von Baer who, three decades prior to Darwin's publication of *On the Origin of Species*, published a set of 'laws' of comparative embryology in 1828. Of course, we know that there are few laws in biology, so it is better to think of frequently occurring patterns than universally applicable laws. Bearing this in mind, and also restricting our attention to von Baer's first law, which is the fundamental one (the others flow from it), we can state von Baer's main contribution to comparative embryology as follows:

The general features of a large group of animals appear earlier in the embryo than the special features.

If this is true, it means that the earliest embryonic stages are the most similar in any cross-species comparison; and the embryos then progressively diverge so that the later embryonic stages are less similar (and, though it is sometimes not explicitly stated, the post-embryonic stages are less similar again). This pattern has been referred to as von Baerian divergence (or sometimes deviation) and is illustrated in the context of a three-species comparison in Figure 17.2.

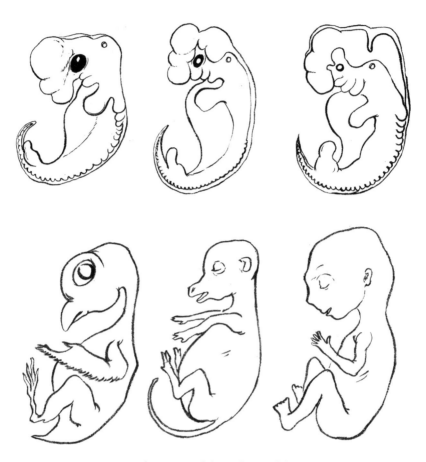

FIGURE 17.2 Von Baerian divergence of the embryos of three vertebrates: a hen, a cow and a human. Notice that, at the earlier of the two stages shown, the embryos are all rather similar. At the second stage shown, it is apparent that we have a bird, a hoofed mammal and a primate. In later developmental stages (not shown), the actual species can be recognized.

We have seen that this is a pattern, not a law; and we have noted that it often will not apply, at least in a straightforward way, to animals that develop via a larval stage. But is it really correct, even with those two restrictions? The answer to this question is 'no', unless we make a further restriction – to what might be called the mid and late stages of embryogenesis, which are actually the ones shown in Figure 17.2. The earliest stages are often *more* different from each other than are the middle ones; this fact means that we often see, in cross-species comparisons, a pattern of convergence-and-then-divergence rather than

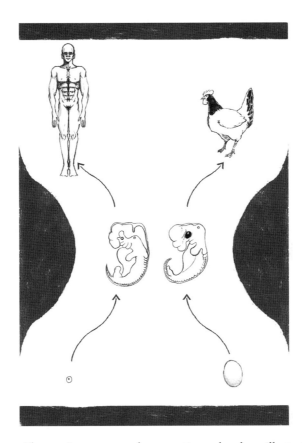

FIGURE 17.3 The egg-timer pattern of comparative embryology, illustrated by a comparison of human and hen. This is a variant on the pattern suggested in the nineteenth century by von Baer. He described a pattern in which the embryos of different species diverged from each other over developmental time. However, if the very earliest stages of development are included, we see a pattern of convergence followed by divergence that can be likened to the shape of an egg-timer or hourglass.

divergence all the way through. This more complex pattern has been referred to as the developmental egg-timer or hourglass (Figure 17.3; 'egg-timer' was first introduced in this context by the Swiss biologist Denis Duboule in 1994). This name reflects the fact that often it is not the very earliest developmental stages that are the most similar, but rather the somewhat later stages that give the constriction or waist that characterizes the shape of an egg-timer.

We now have some questions to ask about causality. Why do we often find this egg-timer pattern? That overall question can in turn be split up into smaller, component ones. Why the initial difference? Why the convergence to a narrow waist? And why the subsequent divergence?

Before proceeding to the answers to these questions, let's get rid of the term 'waist'. There's a scientific name for it, coined by the German embryologist Klaus Sander in 1983: the *phylotypic stage*. This name is intended to imply that the stage concerned is conserved throughout most or all of the relevant phylum or subphylum. In insects it is the segmented germ-band stage; in vertebrates it is the stage often referred to as the pharyngula. Both of these are post-gastrulation.

I'm now going to take the above three related questions in the opposite order to that in which they occur in developmental time (thus starting with the last one). This reflects the way they have been dealt with in the history of scientific thought on this subject.

Divergence in the development of two or more kinds of animal after the phylotypic stage is a consequence of standard Darwinian natural selection. So, for example, the gill slits of fish pharyngulas will go on to become gills, while those of humans will not. The natural selection that eliminated gills in terrestrial vertebrates that have evolved lungs does not result in a pattern of gills-until-hatching followed by the sudden conversion of gills into lungs in hatchlings, be they lizards, hawks or humans. Instead, the relevant developmental trajectories got modified all the way from the phylotypic stage to the adult.

It's not surprising that this kind of evolutionary modification penetrates back beyond hatching/birth, into the embryo. Natural selection works with whatever variation is available – more on this in Chapters 22 and 27. Sometimes, the nature of the variation is such that the developmental trajectory gets modified near its start; in other cases the variation is different and developmental trajectories do not begin to diverge until somewhat later. This is just Darwinian selection working in its usual messy way, memorably described by the French biologist François Jacob, in a paper published in *Science* in 1977, as evolution being a 'tinkerer' rather than a designer-from-scratch.

The thing that is surprising, then, is that there appears to be a sort of early buffer that gets hit if evolution 'tries' to modify the developmental trajectory back to, or beyond, a certain point – the phylotypic stage. This stage seems to be impervious to change, or nearly so. Why? The consensus view is that tinkering with the phylotypic stage will almost always wreck the integration of the embryo and make downstream development inviable.

But if this is true, why are there different very early developmental processes that converge towards the phylotypic stage? The answer is thought to lie in the necessity of these very early stages in a sense adapting to what we might call the geography of the egg, including the location and amount of yolk. Hence the flattened blastodisc of birds, sitting atop a large yolk mass, in contrast to the more spherical blastocyst of mammals.

These explanations of the different components of the egg-timer pattern seem plausible; and they may even be correct. Yet we should return to the pattern itself and inspect it more carefully because it, like other evolutionary patterns, is a messy one and we should not make the mistake of thinking of it as neat and tidy. The Swiss biologist Denis Duboule, who came up with the egg-timer term, emphasized that the point of constriction was not really a clearly defined stage occurring at a specific point in developmental time, but rather a series of stages. The Leiden-based English embryologist Michael Richardson made a similar point in 1995 when he stressed that we should think not in terms of a phylotypic stage but rather of a phylotypic period.

So we have reached an important point in our understanding of comparative embryology: we can recognize a pattern (the egg-timer); we know that it is a general one; we can come up with a causal explanation for it; and yet we can also acknowledge the pattern's messiness in the sense of there not being as clearly defined a phylotypic stage as was first thought. This is all very positive. But our story of comparative embryology is not yet complete. We have still to look at the work of the famous nineteenth-century embryologist Ernst Haeckel, and the comparative embryological pattern that is often associated with his name: recapitulation. Both he and it have been, and continue to be, controversial.

The term *recapitulation* is used for situations in which animals of a descendant species pass through developmental stages that resemble an ancestral species in some way. There is no doubt that this frequently happens. But despite this fact, recapitulation has often been criticized and has been thought by some biologists to be a disreputable idea. We need to examine the reasons for this strange state of affairs.

In fact, von Baer's laws, of which the first one was given earlier, were laid down partly to describe what is actually seen when embryos of different animals are compared, but partly also to state in a forthright way what is not seen. In relation to this latter point, von Baer was specifically trying to refute earlier suggestions of something akin to what I have described above as recapitulation. However, these early suggestions were made mainly by members of the pre-Darwinian late-eighteenth-century school of thought known as nature-philosophy; they were made not in evolutionary terms but in idealistic ones related to supposedly higher and lower beings. Also, some proponents of this initial version of recapitulation envisaged higher animals going through developmental stages that resembled the *adult* forms of lower ones, which is not the case. Von Baer's fourth law was explicitly aimed at dismissing such a notion. He stated: "Fundamentally, therefore, the embryo of a higher animal is never like [the adult of] a lower animal, but only like its embryo." This translation from the original German is taken from Stephen J Gould's 1977 book *Ontogeny and Phylogeny*.

Haeckel published his major treatise, *Generelle Morphologie*, in 1866, just a few years after Darwin's publication of *On the Origin of Species*. He accepted evolution and came up with a proposed evolutionary law: ontogeny recapitulates phylogeny (or, development repeats evolution). This is often referred to as the biogenetic law. Haeckel has been criticized because this law can at face value look like a return to the pre-evolutionary version of recapitulation that von Baer was keen to refute. However, a careful look at what Haeckel actually said shows that this interpretation of his biogenetic law may be wrong.

Here is what Haeckel says in *The Evolution of Man* (an 1896 translation of his German book *Anthropogenie*); try to ignore the use of 'evolution' to mean 'development' in the quote, a usage that was common in the late nineteenth century:

> The fact is that an examination of the human embryo in the third or fourth week of its evolution [= development] shows it to be altogether different from the fully developed Man, and that it exactly corresponds to the *undeveloped embryo-form* presented by the Ape, the Dog, the Rabbit, and other Mammals, at the same stage of their Ontogeny [= development]. (The italics and the words in square brackets are mine.)

This makes it seem that Haeckel's recapitulation was more akin to the views of von Baer than to those of the earlier nature-philosophers, and indeed Haeckel frequently praises von Baer's work. It appears from this quote that Haeckel did not see human embryos as passing through stages resembling adult dogs, for example; he saw them as resembling dog embryos, which they do. His main mistakes were to use 'exactly' to describe the resemblance between human and other mammalian embryos; and to try to raise recapitulation to the status of a law rather than just a frequently occurring pattern.

It would seem from the above that there is no real conflict between the main thrust of von Baer's and Haeckel's views of comparative embryology, in the following sense. Human embryos having gill slits can be seen as part of a pattern of von Baerian divergence, because fish embryos also have gill slits; but it can also be seen as Haeckelian recapitulation, because human embryos exhibit this character (gill slits) and so did the last common ancestor of humans and fish. So, in our development, we *diverge* from a present-day fish species but we also *recapitulate* a feature of our fishy ancestors.

There is, however, a catch. Parts of Haeckel's 1866 *magnum opus* seem to indicate that he, like the earlier nature-philosophers, thought of recapitulation as involving the adult stages of ancestors in the embryos of descendants. There is thus a question mark about this issue. My personal view is that Haeckel was too good an embryologist to think, for example, that at a certain stage in its development, a human embryo looks like an adult fish, for example a pike or a salmon. I do not even believe that Haeckel thought there was a stage in human embryogenesis that exactly resembled a *generalized* adult fish.

A wonderfully wide-ranging book by Haeckel, *The Riddle of the Universe*, appeared in 1900. Chapter 4 of this book is entitled "Our embryonic development". At the start of this chapter, Haeckel says: "I will confine myself here to a brief survey and discussion of the most important phenomena." Having said that, he goes on to discuss recapitulation and the biogenetic law both in chapter 4 and in chapter 6. He compares human embryos with the embryos of other animals, but nowhere in the book does he say that humans pass through stages resembling *adults* of other species.

After Haeckel's major works, there was a long gap (roughly 1900–1980) in which comparative embryology was unpopular; during this period evolutionary and developmental biology parted company and remained largely separate endeavours. But they came back together again in the early 1980s in a very productive way. In the past few years we have come to recognize an egg-timer pattern in the activity of the genes controlling development, as well as in development itself. Indeed, we have come to learn all sorts of things about the genetic control of development, and how it exhibits both commonalities and differences among different kinds of animal. This new dimension to an old story will be the subject of Chapter 19.

18   Larvae, mouthparts and moulting

In the previous chapter we dwelt largely on vertebrates. The embryos shown to illustrate patterns such as von Baerian divergence or the developmental egg-timer all belonged to one vertebrate group or another. Also, when discussing the complications to such patterns provided by indirect development and larvae, the larvae we examined were those of our reasonably close relatives – the echinoderms or spiny-skins, and specifically sea urchins. What this means is that the examples that came into focus in Chapter 17 were all from the mouth-second super-group of animals, the deuterostomes.

In this chapter we will make good our omission from the previous one of the complementary group of animals, the protostomes. This super-group in fact includes the bulk of the animal kingdom, both in terms of number of phyla – it has about 20 of the animal kingdom's 35 or so phyla – and (because the arthropods are included here) the largest aggregate number of species. Given its size, we need to probe a bit into its internal structure, in the sense of the evolutionary relationships among its constituent phyla. But, as ever in this book, we need to do this without introducing a whole new raft of technical terms. This is a difficult challenge but hopefully not an insurmountable one.

I mentioned earlier that there has been a radical reappraisal of our views on high-level animal relationships in recent times. But this reappraisal has not been evenly spread through the kingdom. It is our views on the relationships among different protostome phyla that have seen the biggest change. This change can be explained by looking just at four of the roughly 20 phyla involved; and it makes sense to focus on four that we have already dealt with so that new phylum names do not have to be introduced here. (However, a list of phylum names can be found in the Appendix, for those who are interested.)

The four phyla we will focus on are: the arthropods, the molluscs, and two groups of worms – the annelids (segmented worms) and the nematodes (unsegmented roundworms). Three of these have already had a chapter, or most of a chapter, devoted to them and the fourth (annelids) have popped up here and there.

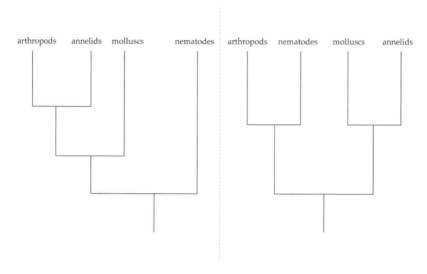

FIGURE 18.1 Two cladograms: old and new views of the pattern of relationship among four phyla of protostome animals. The old view (left) was based on the presence/absence of segmentation and of the body cavity known as the coelom. The new view (right) is based on molecular characters and also on whether the animals concerned have: (a) a form of growth characterized by moulting, or (b) either a larval stage called a trochophore or a circle of tentacles around the mouth called a lophophore.

One view of the evolutionary relationships among these groups that was popular for several decades, up to the 1990s, is shown in Figure 18.1. It is based on information in the 1964 book *Dynamics in Metazoan Evolution*, by the English zoologist R. B. Clark, which was referred to in Chapter 8, and several other sources from the same era. (By the way, Metazoan really just means animal. It's a hangover from the era in which some single-celled creatures were considered to belong to the animal kingdom; these were called the Protozoa, a term no longer used, while all multi-celled animals were called the Metazoa.) The contrasting modern view of the relationships among our selected four phyla is also shown in the figure. In both cases the patterns are shown in their simplest possible form – cladograms – so there is information about relative closeness of relationship but no information on actual divergence times.

In the mid-twentieth century, all schemes of animal relatedness were based on the structure of the animals concerned – including the structure of their embryos and (where applicable) their larvae. There was no input from comparative molecular data because such data did not exist.

It should be recalled that the structure of DNA was only discovered, by James Watson and Francis Crick, in 1953; and that the basics of gene function, in terms of the manufacture of proteins via RNA and the genetic code, were still being worked out in the late 1950s and 1960s. Many important additions to our picture of the way genes work did not occur until later. For example, the fact that a typical gene has long non-protein-coding stretches within it (introns) did not become apparent until the 1970s (this is more generally the case in animals and other eukaryotes than it is in Bacteria and Archaea).

It was not until the 1980s that there was enough DNA sequence data from a range of different animals for the production of schemes of evolutionary relatedness among animal phyla based either solely on the new molecular data or on a mixture of this and the older morphological information. Publications involving this new approach have been coming out over the last three decades or so, and are continuing apace. One of the most important of these was the one mentioned briefly in Chapter 14 when discussing which groups were evolutionarily closest to the arthropods – authored by Anna Marie Aguinaldo *et al.* and published in the journal *Nature* in 1997.

To understand the basis for the old view of protostome relationships and how it was revolutionized in 1997, we need to examine four morphological/developmental characters. We also need to delve a little deeper into the nature of DNA sequence comparisons than we did when looking at the basis for recognizing protostomes and deuterostomes as animal super-groups in Chapter 16.

Here are the four morphological/developmental characters: segmentation, a type of body cavity called the coelom, growth by moulting, and trochophore larvae. We've met some of these before, but not others; in the following account I'll be brief with the former but less so with the latter.

An animal is described as segmented if it consists of a series of body-segments along its main (i.e. antero-posterior) axis. Of the four phyla we're focusing on, this is the case with arthropods and annelids, but not with molluscs or nematodes. A few molluscs show serial repetition of some organs, but this is not a significant enough feature for the group as a whole to be considered segmented.

The coelom, when present, is the body cavity within which the main organs are suspended – as in the case of the human heart, liver and kidneys. In general, animals can be divided into those with and those without a coelom – called coelomate and acoelomate respectively. However, there are some animals that are in a sense intermediate – these can be referred to as pseudocoelomate. Of our four phyla-in-focus, arthropods, annelids and molluscs all have a true coelom, while nematodes do not (they have been described by many authors as fitting into the pseudocoelomate category).

Before proceeding further, check the information in the last two paragraphs against the first of the two cladograms in Figure 18.1 (the old view of protostome relationships). You'll see that the presence/absence of segmentation and a coelom provided the basis for the pattern of evolutionary relationship depicted. It should therefore come as no surprise that the subtitle of the book by R. B. Clark was *The Origin of the Coelom and Segments*.

Now we turn to growth by moulting. We noted in Chapter 14 that the new view of the position of arthropods in the animal evolutionary tree involved the creation of a new super-group of phyla called the Ecdysozoa, or moulting animals. Nematodes as well as arthropods grow via a series of moults, but molluscs and annelids do not. (Annelid worms grow by adding segments at their posterior ends, but no moulting is involved in this process.)

The presence of trochophore larvae (which were pictured earlier: Figure 12.2) goes with the absence of moulting and so has the opposite pattern of occurrence among the four phyla: molluscs and annelids have these larvae; arthropods and nematodes do not. (A caveat: molluscs and annelids that have invaded the land have lost their larvae.) So, returning to Figure 18.1, but this time looking at the new view in the second cladogram, it can be seen that the evolutionary invention of growth-by-moulting and of trochophore larvae give the pattern shown. However, although it is therefore true that this pattern corresponds to some morphological/developmental characters, the rationale underlying the first suggestion of it in 1997 was based largely on molecular data. So it's now time to examine the way such data are used to generate evolutionary trees.

The starting point for building molecular trees is the same as that involved in building morphological ones: examining degrees of similarity. And in some respects the job is easier for molecular data because the degree of similarity is looked at along a single axis (a DNA strand) rather than being multidimensional, as is the case with morphology. However, we also saw the caveat that it was necessary to choose a gene (or other stretch of DNA) that evolves at an appropriate rate for the depth of the comparisons being made. Thus, for example, attempts to build a molecular tree for Darwin's finches, which have evolutionarily separated from each other quite recently, need a fast-evolving gene, while attempts to build a tree of the relationships among phyla, involving, as they do, lineage splits that occurred in deep time, need a slow-evolving one. Of course, in either case it is desirable to use not just one but several genes that are evolving at the desired rate.

A frequently chosen gene to use for the examination of deep lineage splits is the one making a particular kind of RNA – the kind called 18S rRNA. This requires some explanation. The r stands for ribosomal, because this kind of RNA makes up the bulk of the small intracellular bodies called ribosomes (Figure 18.2). The 18S distinguishes one

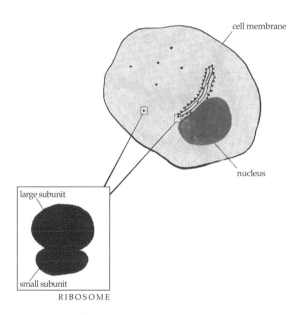

cell membrane

nucleus

large subunit

small subunit

RIBOSOME

FIGURE 18.2 Ribosomes. These are little structures found inside almost all cells, both of animals and of life-forms belonging to other kingdoms. They are the sites within cells at which messenger RNA is translated into protein. Some are attached to a membrane, others are free in the cytoplasm. Each consists of a large and a small subunit. The 18S RNA is found in the small subunit. (For more information about gene function, including the role of messenger RNA in protein synthesis, see next chapter.)

particular kind of rRNA molecule from another on the basis of size – there are also other sizes, notably 5S and 28S. We don't need to delve into these codes too deeply. We just need to note that the numbers are a measure of the size of the molecules, so that of the three numbers given the 5S molecule is the smallest and the 28S the largest. (The S comes from the name of the Swedish chemist Theodor Svedberg.)

The gene that makes 18S rRNA is slow-evolving. Studies on its sequence in many animals (starting with the 1997 paper by Aguinaldo *et al.*) have suggested that there are two main groups of phyla within the Protostomia: the moulting group now called the Ecdysozoa and another group, not mentioned before, with the formidable name Lophotrochozoa. This long word is a fusion of three shorter ones: lophophore, trochophore, and zoa, the last of these simply meaning animals. Lophophores are feeding structures with tentacles around the mouth. These structures have not previously been discussed here because they aren't found in any of our four phyla-in-focus, but it turns out that they are found in several related phyla. So, members of the group Lophotrochozoa have either lophophore feeding structures or trochophore larvae.

The validity of the two groups of phyla – Ecdysozoa and Lophotrochozoa – has been confirmed by studies on other slow-evolving genes in addition to the 18S rRNA gene. So there is now a consensus on this issue. And it is an important one, because the lineage split that led to these two groups is one of the deepest splits in the animal kingdom. It's not quite as deep as the split that produced the protostomes and deuterostomes, of course: that would be logically impossible. But it's the next deepest major split after that one. Recognition of it leads to a perceived structure of the main part of the animal kingdom – the Bilateria – that is essentially an expanded form of the second cladogram in Figure 18.1; it is shown in Figure 18.3.

This is as far as we are going to go in this book in terms of the large-scale structure of the animal kingdom, that is, its structure above the level of the phylum. Zoologists who conduct research in this field are still hard at work, both testing what we think we know already, such as the existence of the super-groups of phyla that have been our focus in this chapter, and also extending knowledge into corners of the animal kingdom where we still admit that we don't really know what the correct pattern of relatedness is, especially in relation to some of the so-called minor phyla. There may yet be surprises ahead, but my guess is

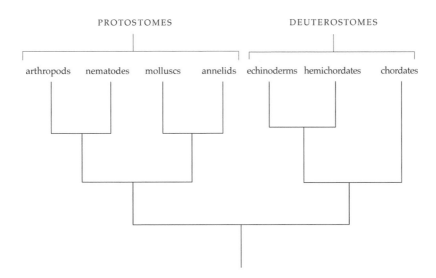

FIGURE 18.3 An expanded cladogram based on the new view of protostome relationships. This one incorporates the right-hand panel of Figure 18.1, but also selected deuterostome phyla.

that we're a lot closer to the true animal tree than we were just a few decades ago, and that the magnitude of reappraisals will decrease over time.

We should now briefly discuss *how* the lineage splitting and evolutionary modification that led to animals belonging to different high-level groups such as the Ecdysozoa and Lophotrochozoa took place mechanistically. These things didn't happen on lines of descent drawn on a piece of paper – rather, they happened in real, albeit long-gone environments. They involved the Darwinian natural selection and reproductive isolation discussed in Chapter 5. But that's too vague a statement to be of much use. The evolutionary radiation of Darwin's finches over a mere few million years involved those processes too.

This point brings us back to a recurring question: is evolution scale-dependent? Or, to put it another way, do the deep splits that produced the major branches of the animal evolutionary tree arise simply from many small-scale adaptive radiations compounded over time, with the resultant branches sometimes becoming more distinct from each other because of extinction? This, in my view, is one of the most important unanswered questions in the whole of evolutionary biology today. We'll take a quick look at it here and then return to it in later chapters.

Let's take the deep splits discussed in this chapter and Chapter 16 together. In other words, let's include both the split between the biggest super-groups in the animal kingdom (protostomes and deuterostomes) and the split within the protostomes that led to its two constituent groups of phyla, the Ecdysozoa and the Lophotrochozoa.

We've now seen the main changes involved in these two splits. The first split may have been characterized by a reversal of the order of formation of the mouth and anus, though, as we saw in Chapter 16, the patterns of formation of the gut in protostomes and deuterostomes have turned out to be more complex than was initially thought. One way to think about possible forms the protostome–deuterostome split may have taken is to look again at that enigmatic animal the urbilaterian – that is, the original bilaterally symmetrical animal that lived many millions of years ago, probably in the Ediacaran period. Recall that we have no fossil evidence of this creature and that there are competing hypotheses of what it was like: the simple-urbilaterian and complex-urbilaterian views. When discussing these earlier (Chapter 8), we focused in particular on the question of whether the urbilaterian animal was segmented or not. However, we can also ask: did it have a through-gut with an opening at each end, namely a mouth and an anus?

As noted when we discussed the phylum Cnidaria (jellyfish and their allies: Chapter 6), this basally branching group of animals has a blind-ending digestive cavity with a single opening that functions as both mouth and anus. Suppose for a moment that the urbilaterian had a similar arrangement. If so, then, as we noted in Chapter 16, there may have been no reversal of mouth and anus formation in the protostome–deuterostome split, but rather independent invention of a through-gut in the stem lineage of each of the two groups, involving one group accidentally hitting on the mouth-first solution, the other accidentally hitting on the mouth-second solution.

It's worth examining what is meant by the use of 'accidental' here. Recall that evolution works from a starting point of the existence of variation, which can then be acted upon by natural selection. Some kinds of variation are always present in natural populations of pretty much all animal species – for example, variation in adult body size. But other kinds of variation are not. In the present context, we can ask whether there are any animal species that have variation in the order of formation of the mouth and anus. As far as I am aware there are none. This means that in a (hypothetical) urbilaterian species with a blind-ending gut, it might be the case that in one of its populations living in one part of the world a variant arose in which the developing gut broke through at the other side of the embryo from the one at which it started, and became the anus. And in another population living somewhere else a variant may have arisen in which the same breaking through happened but the embryonically later opening became the mouth.

Now, turning to the second great split in the bilaterian animals, the within-protostome split that led to the origin of the Ecdysozoa and the Lophotrochozoa, what changes took place here? As we've seen, they include the origin of growth-by-moulting in one group and of trochophore larvae in the other.

Were these various origins adaptive? A case can easily be made that the evolution of a tough and perhaps predator-resistant cuticle that brings with it a need to grow by moulting would have been adaptive. Also, a case can be made that the evolution of a particular type of dispersive larva would have been adaptive, especially in a coastal marine animal species whose adults are not very mobile. Such larvae would help in the geographic spread of the species and also help to prevent inbreeding. And again, a case can be made that the evolution of a through-gut, allowing one-way transit of food and waste material, and thus more efficient processing of resources, would have been adaptive. However, it's less clear why a reversal in the order of formation of mouth and anus – if indeed this happened – would confer a selective advantage.

Our line of reasoning is taking us in a particular direction. It's taking us to a view that perhaps many of the changes involved in deep splits were adaptive, just as were the changes involved in shallow ones – like the splits between different lineages of Darwin's finches. But there are two important differences, as follows.

First, an evolutionary change involving the production of a through-gut from a blind-ending one is not an adaptation to any particular environment. It may be even more advantageous in some ecological contexts than others, but it's perhaps better thought of as a *generalized improvement* rather than as an environment-specific adaptation. (We'll probe this idea further in Chapter 26.) Second, the variation upon which it was based is not of the type that is routinely present in natural populations of any kind of animal. So this sort of evolutionary change is based on 'waiting' (not intentionally, of course) for an unusual variant to arise.

The same rationale as applied to the origin of a through-gut in the above paragraph can also be applied to the evolutionary origin of tough cuticles and dispersive larvae. Even though there is a definite restriction on the advantage conferred by a trochophore larva, in that it is limited to aquatic environments (and so has been lost in lineages invading the land), it is still advantageous in a very wide range of aquatic environments rather than just a few of them. This is rather different from the evolution of a particular size and shape of beak in a species of Darwin's finch, which may, at least in some cases, only be adaptive on a single Galapagos island with its unique array of food items.

So the two take-away messages involving deep splits in the animal kingdom are: (a) that the changes involved are generalized improvements rather than narrowly environment-specific adaptations; and (b) that they are based on 'waiting' for unusual variants to arise rather than on routinely occurring, omnipresent variation, such as that in body size. We will consider these important messages further in later chapters. At this stage, it is best to regard them as interesting and provocative suggestions rather than as firm conclusions.

19 The animal toolkit

We've now got a pretty good idea of the overall structure of the animal kingdom. We know that a small number of early lineage splits produced sponges, cnidarians, bilaterians, and a few smaller groups. We know that there was a deep split within the Bilateria, producing the groups that I have informally called the mouth-first and mouth-second ones, these being translations of their official names, protostomes and deuterostomes. We also know that phyla are very asymmetrically divided between these two groups, with the majority belonging to the protostome one. And we have seen that this group underwent a deep split into the moulting animals (Ecdysozoa) and the non-moulting ones that are collected together under the unwieldy name of Lophotrochozoa, because most of them have either a trochophore larva or a ring-of-tentacles (lophophore) feeding structure. Finally, we know where the three biggest animal phyla – the arthropods, molluscs and chordates – fall within this overall pattern of relationships.

Here's a distinction that we haven't yet made but that we should make now: the difference between diversity and disparity. I've labelled the arthropods, molluscs and chordates as the 'biggest' animal phyla because of their number of species. The Arthropoda, with more than a million species, is the biggest phylum by far. Of its four subphyla, the Insecta is easily the largest, containing about 80% of arthropods. So the Insecta has a high diversity – meaning a large number of species. But it's quite low in disparity.

What does this last statement mean? What, in the context of the animal kingdom, is disparity? It's a term that was championed by Stephen J. Gould; it refers to differences in animal form. We can see how it differs from diversity by comparing the insects with their sister-group, the crustaceans. The latter have only about 10% as many species as the insects, yet they exhibit a much wider range of body forms and hence are more disparate. For example, all adult insects have three pairs of legs, but adult crustaceans can have anything from zero to more than 30 pairs.

If we now move out again from arthropods to the whole animal kingdom, one of the most striking things about our kingdom is its

incredible disparity in overall body form. Think of representatives of
some of the main branches: jellyfish, snails, starfish and whales. Even
contemplating a few examples, such as these, reminds us that skeletons
may be absent, internal or external; that body length can be about a
millimetre or about 10 metres – a range of four orders of magnitude. The
number of cells in an adult animal can be less than a thousand, as we
saw with the 'elegant roundworm', or many trillions, as in humans.

This great disparity in body form would suggest that the way in which
the developmental process works must, in some ways, be different
among different kinds of animal. And indeed, in *some* ways, it is: in
previous chapters, we've noted different types of embryonic processes
such as cleavage and gastrulation; we've also noted that some animals
have direct development, others indirect – getting to the adult via a very
different sort of creature called a larva. But in some other ways the
developmental processes of different kinds of animal are rather similar.
This is especially true of the genes that are involved in animal development, including both the similarity of the genes themselves (measured
in terms of their DNA sequences) and the similarity of their developmental roles – that is, what the gene products do in terms of building the
animal from the starting point of a fertilized egg.

The discovery of this kind of developmental similarity underlying the
incredible disparity of animal form, which was made (or at least started)
in the 1980s, must rank as one of the greatest biological discoveries to
date. Yet despite this fact, the names of the people involved are not
familiar to the general public in the same way as are the names of the
discoverers of the structure of DNA in 1953, James Watson and Francis
Crick. Here are two of them, so you can check if I'm right in your case:
Matt Scott and Bill McGinnis (I'll provide some other names later).

Having mentioned the structure of DNA, I should clarify how the
1980s discovery differed from that 1950s one. All animals, and indeed
almost all life-forms (some viruses are exceptions) have DNA as their
genetic material. To put it another way, all our genes are made of DNA,
albeit this is packaged up with, and wound around, protein molecules.
The DNA of the genes of all animals has the same structure – the
famous double helix discovered by Watson and Crick. The two helices
are essentially the same throughout the animal kingdom – and beyond.
The way in which genes differ is not in the helices, which have a
standard structure of alternating sugar and phosphate units, but in the

links between them. (If this fact is already familiar to you, skip the next three paragraphs, which provide an introduction to 'DNA basics'.)

To obtain a simple mental image of these important links between the DNA helices, we can ignore the twists involved in the helices – we can think instead of two long straight-line structures that are joined by many shorter lines; the best analogy is a ladder. We can call the long lines 'strands' and the shorter ones (the rungs) 'links'. What makes any particular gene unique is the nature of its links. Each link involves a connection between a projection from one strand and a corresponding projection from the other. These projections are called bases (because from a chemist's perspective they are the opposite of acidic, which is *basic* – but that's of no more interest to us than the fact that the A of DNA stands for acid). There are four of these bases and they're usually known by the initial letters of their names rather than those names in full. In alphabetic order, they are: A, C, G and T (or, for those who prefer names to initials, adenine, cytosine, guanine and thymine). Because of their precise structure, A binds to T, and C binds to G (Figure 19.1).

Each gene is unique because it has a particular sequence of these bases along its length. For example, at the start of one gene we might find that the sequence of the first six bases is AATCCG, while in another it might be ACGTTT. A typical gene is many thousands of bases long, so the probability of any two genes being identical, whether different genes in the same animal or the equivalent gene in different animals, is effectively zero. But comparing genes can reveal varying degrees of similarity in their sequence, from a small percentage up to 100% (comparing the same gene in identical twins). Two genes will never have 0% similarity because, with only four bases available, two randomly chosen genes will end up with the same base at many positions along their length just by chance.

Of course, since the DNA ladder has two strands, a geneticist can look at the base sequence along one of these, or along the other, complementary strand. Usually, in terms of the function of a particular gene, one of the strands is the sense strand (and the other is called the anti-sense strand), but some genes are now known in which one protein product can be made from one strand and a different protein from the other. However, if we ignore this complexity and assume that it's OK to choose just one strand for sequence comparisons, then it's not so hard in principle to generate a percentage value for the sequence similarity of

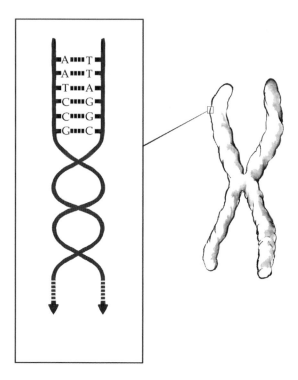

FIGURE 19.1 Part of a gene, showing the way in which complementary bases bind to each other, thus forming links between the two strands of DNA, turning it into a sort of ladder; this ladder is not flat, but rather coiled, hence the famous double-helix structure of DNA. This helix is supercoiled several times and wrapped around protein molecules, thus producing the chromosomes (right) that can be seen with an ordinary microscope.

any two genes. There are lots of problems in practice, such as different gene lengths, but there are various ways of overcoming these.

There are many different types of gene and many ways of classifying them. For our purposes here I just want to focus on two categories: housekeeping genes and developmental genes. And I'll start with a warning that while many genes fall into one or other of these categories, some genes fall into both and some into neither.

Housekeeping genes are so called because they make the proteins that are part of the general housekeeping activities of pretty much any cell. For example, cells need energy. They get a lot of this from metabolizing glucose. This metabolism involves lots of enzymes working in sequence in what is called a metabolic pathway. The first step in this particular pathway (which has the overall name of glycolysis) is catalysed by an enzyme called hexokinase. This is thus a housekeeping protein, and the gene that makes it is a housekeeping gene. The above is true of almost all cells in all animals.

Developmental genes, in contrast, make specific things happen in particular parts of the embryo (or a later developmental stage) at particular times. Thus by definition they should be expressed – i.e. switched on and making their protein products – only in some cells. There is a developmental gene in humans, and in vertebrates generally, that makes a protein called Sonic hedgehog (we'll get to the reason for this odd name shortly), one of whose jobs is to tell certain cells in the developing nervous system to become motor neurons. There's a complication because many developmental genes have two or more distinct roles, but nevertheless this does not affect the fact that they need to be expressed only in certain cells and at certain times, in contrast to housekeeping genes, which are typically expressed in nearly all cells for much of the time.

One of the difficulties in categorizing genes is that many authors talk about 'regulatory genes' – referring to those genes that control the activities of others. The relationship between this term and my two above categories – housekeeping and developmental – is not as simple as it might at first seem. Some authors equate regulatory and developmental genes because, almost by definition, the genes that control development must be able to regulate the activities of other genes. However, housekeeping genes need to be regulated too; they can't just be switched on at full throttle all the time. The genes that regulate the expression of housekeeping genes are regulatory, again by definition, but they are not developmental. Therefore, I will avoid the term regulatory gene because I think it can be confusing. I will also not deal further with genes that serve a purely housekeeping function, as these are not directly relevant to development. All that follows, then, concerns developmental genes.

How does the discovery of DNA structure in the 1950s relate to the later (1980s) discovery that much of the genetic control of development is shared by many animals of very disparate body forms? To deal with this question, let's put ourselves back in the year 1978: after one discovery but before the other.

It was already clear in 1978 that there was great uniformity at the DNA level, underlying the great diversity of the living world at the morphological level. But what was not known was how far 'up' towards morphology the uniformity extended before it disintegrated and was replaced by variation. This may seem a strange way to put it, but hopefully what I mean by 'up' in this context will become clear shortly.

Let's start with the DNA double helix: it was clear in 1978 that this was essentially the same everywhere. Now, if we move 'up' to gene function – the making of proteins via the intermediary molecule of messenger RNA – this general process also is uniform through the animal kingdom and beyond. The same genetic code operates for nuclear genes in all animals. It's a triplet code in the sense that every three consecutive bases of DNA code for a single amino acid; and proteins are just strings of amino acids, albeit they become rather elaborately folded.

A major discovery in 1977 revealed that gene function was more complicated than had been thought because, rather than the code being read along the full length of the gene, there are sections of the gene (called introns) the RNA corresponding to which gets chopped out of what will become messenger RNA and discarded. However, while the extent to which this occurs varies considerably, being much rarer in bacterial genes than in those of animals, plants or fungi, it is generally true of animal genes – and so, within the animal kingdom, it still counts as uniformity rather than diversity of mechanism.

The molecular processes described so far are in common to both housekeeping and developmental genes. But to progress further 'up' towards the building of animal form, we now need to focus on the latter. And the question arises at this point of what developmental gene products actually do. In fact, most of them do one of two things.

Some of these products act within a cell nucleus to turn one or more other genes on and off, or to more subtly regulate their rate of making their product. These 'turning on/off' proteins are called transcription factors, for the simple reason that the process of gene expression can be thought of as a two-stage one, with the first stage (making messenger RNA) being called transcription, the second (making protein from the mRNA) translation. A transcription-factor protein made by one gene acts to regulate the transcription of another gene.

Other developmental gene products are the components of what are called signalling pathways – interlinked series of protein–protein inter-actions that send signals from one cell to another, and indeed also from one part of a cell to another. Inasmuch as a signalling pathway can be thought to have a start-point, it is the secretion of a protein out of one cell; this secreted protein (called a ligand) will attach to receptor proteins on the outside of other cells, thus starting a sort of chain reaction within those cells, often ending with certain transcription factors attaching to their target genes and switching them on or off.

As can be seen from the above account, transcription factors can be thought of as the end of a signalling pathway – though of course in development there is no real end (or start) because those transcription factors will go on to do something else. But we need some way of simplifying things in order to dissect how the appallingly complex overall process of development works – and this approach is as good as any. So, for now, our simplified view of developmental genes is that they make proteins that interact in the following way: $L \rightarrow R \rightarrow I \rightarrow T$; where L stands for ligand, R for receptor, I for intermediate and T for transcription factor (Figure 19.2).

I want to focus attention on a small number of ligands and transcription factors – coincidentally connected by the fact that their names all begin with the letter H. But before jumping into the detail of these few examples, we need a little more rationale in terms of that 1978 view of things, when DNA structure was known about but developmental genetics was in its infancy and signalling pathways were largely unknown, as were the identities of most developmental genes.

Picture biologists back in 1978 trying to think about the potential similarity/difference of developmental genes doing similar things but in different animals. The prevailing view among these folk was that the genes concerned would be different. For example, the structures of an insect eye and a vertebrate eye are entirely different; so it seemed logical that their development, and the genes that controlled it, would be different too. In fact, it seemed likely that they would be just as different as genes in the same animal doing very different jobs – for example those in humans that produce the proteins haemoglobin (for oxygen transport in the blood) and trypsin (a digestive enzyme in the intestine).

The revolution in thinking about this issue that began in the 1980s led to a situation where biologists of the twenty-first century realize that in fact the genes making eyes or other body parts (e.g. legs) in different animals are similar, not different, despite the major differences in the structures concerned – think of a fly's leg and a human one, which are just as fundamentally different in structure as our eyes and a fly's eyes

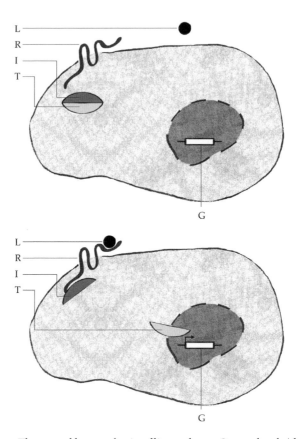

FIGURE 19.2 The general layout of a signalling pathway. One molecule (the ligand, L) binds to another (a receptor, R, spanning the cell membrane). This binding causes a third type of molecule (an intermediate, I, in the cell cytoplasm; there are usually several of these) to do something that ultimately causes transcription factors (T) to switch on (right-angled arrow) one or more target genes (just one here; G) in the nucleus.

are. This revolution in our view of the genetic control of development in different animals was profound. It led to the coining of the term *the genetic toolkit* for the genes-in-common among many animals that are responsible for building the body via the developmental process.

Now for a couple of questions about this toolkit. First, how big is it, in the sense of how many genes does it include? The answer depends on whether we want to include only the major developmental genes, the ones that make big developmental decisions such as which end of an animal will be anterior and where to start making a leg, or whether we also want to include those with much smaller effects – such as those that refine the exact shape of a developing structure. In the former case, the number is 'a few hundred' (admittedly just a rough estimate), or, in percentage terms, about 2–3% of the genes in the genome, since a typical animal genome includes about 20,000 genes. In the latter case, the number is larger, probably a lot larger, though it's hard to be more specific than that.

Second, who has championed the 'genetic toolkit' term and is it a good one? Well, many biologists have championed it, and many more have used it, but one notable figure is the American developmental geneticist Sean Carroll – for example in the 2005 book *From DNA to Diversity* (by Carroll and his co-authors Jennifer Grenier and Scott Weatherbee). I think that the term they are championing is indeed a good one, but that it could perhaps be modified as follows. If anything in an animal cell looks like a tool, it is not a gene but a protein. Genes remain in the cell's nucleus, their main job being to code for things. Those things are proteins, which are mobile agents that go to particular places and do particular jobs. So maybe the 'protein toolkit' for development would be better. Then again, you could call it not after its components/tools but after the things it is used to build – animals; hence my use of *The animal toolkit* for the title of this chapter.

Now to those examples – the ones whose names begin with H. We'll start with Hedgehog – a family of developmental genes, not (at least in the present case) a prickly mammal, though as we'll see shortly there is a link between the two. A search for the genes that control the development of the model fly *Drosophila* was published in 1980 by two German biologists, Christiane Nüsslein-Volhard and Eric Wieschaus – who ended up sharing a Nobel Prize with American geneticist Ed Lewis, who had done pioneering work on *Drosophila* developmental genetics in the 1960s and 1970s.

One of the many developmental genes discovered by the German duo was *hedgehog*. (As an aside here, gene names are always written in italics; the names of their protein products and the names of the gene-groups to which individual genes belong are not.) The method of discovery was to induce mutations in the genes, to observe the nature of the mutant effects on the animal, and thereby to infer the function of the normal (i.e. non-mutated) version of the gene and its product. Fly larvae that had mutant versions of the *hedgehog* gene were prickly, like a hedgehog, in contrast to normal larvae (Figure 19.3).

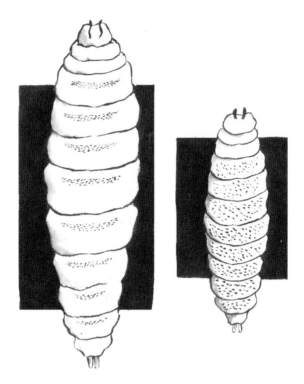

FIGURE 19.3 Normal (left) and mutant (right) larvae of the model fly *Drosophila melanogaster*, seen from below (ventral view). The most obvious difference here is that the small 'prickles' called denticles on the larval cuticle are organized into a series of stripes in the normal larva, but form a continuous lawn in the mutant one. It was this all-over-prickliness that led to the gene involved being labelled *hedgehog*.

As can be seen from Figure 19.3, the prickles, which are actually small, hard, hook-like structures called denticles, are normally restricted to the anterior part of a larval segment, leaving the posterior part of a segment naked. This is because the anterior and posterior parts of the segment have different developmental identities, which are conferred upon them by an interacting group of genes that includes *hedgehog*. When Hedgehog protein is absent or altered sufficiently that it does not carry out its normal function, these two developmental identities are lost; it's as if both the anterior and the posterior part of the segment think they are anterior, and so they both develop bands of denticles. Thus the denticles appear all over the cuticle: rather than occurring in a series of distinct stripes, they appear as a continuous lawn.

Linking this example back to the earlier general account of the things that the proteins made by developmental genes do, the Hedgehog protein is secreted out of cells – so it's one of those roving ligands that are picked up by receptors on the outside of neighbouring cells and begin a signalling pathway which ends with other genes being switched on or off by transcription factors. (Recall, though, the earlier warning about the use of 'begin' and 'end' in developmental processes.)

There is, as you'll probably have guessed, a link between the fly gene *hedgehog* and the vertebrate gene *sonic hedgehog*. Often, in evolution, genes accidentally duplicate, just as they accidentally mutate. And sometimes after duplication one or more of the duplicated copies duplicates again, resulting in there being three, four, or more copies of the gene concerned. These may then diverge in function over evolutionary time, because one copy is sufficient to carry out the original function. Thus the extra copy (or copies) can be used as raw material for natural selection to act upon, with the result that it ends up functioning in some other way.

This has happened in relation to the *hedgehog* gene in vertebrate evolution. As a result, we humans have three hedgehog-type genes. They have been named *sonic, indian,* and *desert hedgehog* (a sense of humour is often involved in gene naming). They all perform different developmental functions. Also, note that although they are all similar enough in DNA sequence that they are recognized as belonging to the Hedgehog gene family, none of them does the same thing as *hedgehog* in insects. There are indeed segments in vertebrate development, but they are very different to those of a fly larva, and they don't have denticles. So, at one level the function of the insect and vertebrate Hedgehog genes is similar – they are involved in the Hedgehog signalling pathway – but at another level, 'higher up', they do different things, such as *sonic hedgehog* being responsible for making motor neurons in vertebrates, as we saw previously. (This is just one of several roles of Sonic hedgehog in vertebrate development; it is also, for example, involved in limb patterning.)

Now for another example of a developmental genetic entity beginning with H. A crucial step made in 1984 was the discovery of a DNA sequence 180 bases long called the homeobox. It was discovered simultaneously by research groups working in Switzerland and the United States. Bill McGinnis (one of the two names mentioned earlier) was working in Walter Gehring's laboratory in Geneva, while Matt Scott (the other) was working with Amy Weiner in Thomas Kaufman's laboratory in Bloomington, Indiana. They were both looking at the DNA sequences of certain genes in *Drosophila*, and they both found homeobox sequences in these genes.

To explain what the 'certain genes' were, we need to jump a long way back – to the year 1894, and something else beginning with H. It was in that year that the British biologist William Bateson published a book with the rather uninspiring title *Materials for the Study of Variation*. One of the topics that interested Bateson was a phenomenon called homeosis. This is where the development of an animal goes wrong in a particular way, with the result that a structure is not malformed but rather formed in the wrong place. One of the most famous examples of this is the formation of an extra pair of wings in flies (Figure 19.4). The fly at the bottom of the figure can be said to have undergone a homeotic transformation. Such transformations usually have a genetic basis, in which case they are underlain by homeotic mutations, but they can sometimes occur due to extreme environmental conditions prevailing during development.

Now we get into a story involving lots of H-words that can all too easily get tangled up and lead to misinterpretations. Let's see if I can tell it in such a way that this does not happen. Here goes.

The American geneticist Ed Lewis studied the genes that undergo homeotic mutations in *Drosophila*, producing flies like the one that is pictured in Figure 19.4, together with others of a similar kind – right-structure-in-wrong-place – such as flies with legs growing out of their heads where they should have antennae. These genes can be called homeotic genes, though that's a sort of convenient shorthand for genes-that-can-undergo-homeotic-mutations. As a result of the work of Lewis and others, we now know that there are eight homeotic genes in *Drosophila*, and that they occur in a chromosomal cluster. This cluster is split into two sub-clusters in flies (but not in most other insects), and the two sub-clusters are given names. One, with three homeotic genes, is called the Antennapedia Complex (after the legs-for-antennae transformation); the other (with five genes) is called the Bithorax Complex (after the two-pairs-of-wings transformation).

The 'certain genes' that McGinnis, Scott and others were working on in 1984 were homeotic genes from these complexes. That's why, when a

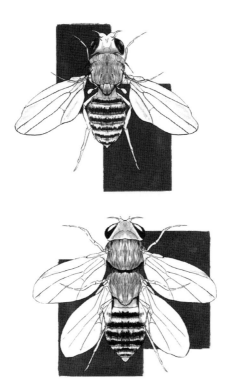

FIGURE 19.4 Normal (top) and mutant (bottom) *Drosophila melanogaster* flies. In the normal form there is a single pair of wings, attached to the middle thoracic segment (the second segment of three in the thorax), and a single pair of small flight-balancing organs (halteres), attached to the third thoracic segment. In the mutant form, the halteres have been transformed into wings, so the bithorax mutant fly has two pairs of wings instead of the normal single pair.

conserved DNA sequence was found in these eight homeotic genes, it was given the name homeobox. ('Box' just refers to the practice of drawing a rectangular box around a sequence of DNA bases that is conserved in the sense that it shows up, with some minor variation, in different genes.) At first it was thought that this sequence was found *only* in homeotic genes. Also, it wasn't clear if it was found in animals that were not flies. However, further studies carried out in the mid-to-late 1980s and subsequently showed that the homeobox was present in many genes in flies (perhaps 200 or so, not just eight) – and indeed in all animals. So, like the Hedgehog gene family, the homeobox-containing gene family represents an important component of the toolkit for animal development.

Here are a few more facts to complete the story – or at least my simple version of it. In any gene containing a homeobox, that 180-base sequence is just a small part of the gene – most animal genes are thousands of bases long. When such a gene is switched on, its protein product includes (again as a small fraction of its total length) a sequence of 60 amino acids called the homeodomain, which is the result of decoding the homeobox. A homeodomain has a DNA-binding role, indicating that the function of the protein of which it is a part is to go into the nucleus and bind to the DNA of target genes, which it can switch on or off. Therefore, homeodomain proteins are transcription factors. Finally (and confusingly!) the term 'Hox genes' is not short for 'homeobox genes'; rather, it's a term used to refer only to the homeotic genes (eight in flies, 39 in humans) – the ones that are found in chromosomal clusters and whose main role in most animals is to specify the correct antero-posterior identities of sections along the main body axis.

Broadening out, the animal toolkit includes many families of genes. We've looked at two of them here – Hedgehog genes and homeobox-containing genes (including the Hox genes). There are many others that we haven't looked at – such as the Pax gene family (including *pax 6*, which has a role in eye formation) and the Dlx family (whose genes are involved in limb formation). If you want more detail, there is plenty of it to be had in more technical books, including the 2005 one by Carroll *et al.* mentioned earlier. The take-home message here is that all animals have a toolkit of many developmental genes and their protein products. Although each animal's toolkit is a bit different from those of other animals, because the constituent genes have diverged somewhat, the *similarities* are very striking, and were unexpected before they began to emerge in the 1980s.

Thus each kind of animal comes to be built, in development, in a way that is part similar, part different, from each other kind of animal, due to this combination of similarity and difference in their genetic toolkits and the ways in which the genes that make them up are deployed. There is much yet to be learned about how any one kind of animal is made, and about the pattern of similarities and differences in the development of any two or more kinds of animals chosen for comparison. So the next three decades may be as exciting as the last three in terms of discoveries in this fascinating field.

20 Vertebrate origins and evolution

Having a special interest in our own evolutionary origins is hard to avoid. As noted in Chapter 16, this interest can be indulged at a variety of levels, and on multiple timescales. For example, at the level of the species we can ask how *Homo sapiens* arose from other species of *Homo* over the last few million years – this is the focus of attention in Chapter 24. At a much higher level, we can ask how the bilaterian animals, of which we are one, arose *hundreds* of millions of years ago (Chapter 8). In between, but closer to the latter than to the former in terms of geological time, we can ask how the vertebrates began.

The earliest vertebrate fossils found so far are from the Cambrian period, and specifically from about 530 million years ago (MYA). As ever, the earliest fossils of a particular group set a latest possible date for the origin of the group concerned, but we would usually expect its actual origin to be earlier. The trouble is that it's hard to do more than guess how much earlier. The first vertebrates may have lived in the very earliest part of the Cambrian – around 540 MYA – but there may have been vertebrates even earlier still, in the Ediacaran period (635–542 MYA).

The closest living relatives of the vertebrates are the non-vertebrate chordates. There are two groups of these, which were pictured in Figure 4.5 – the cephalochordates (lancelets, which superficially resemble small, slim fish) and the urochordates (including the sea squirts, the adults of which are quite bizarre and whose chordate links are only evident from their tadpole larvae). These two subphyla are very different from each other in terms of their body forms – and also in their degree of morphological disparity and in their diversity of species.

There are only about 20 species of extant lancelets, and their body form varies rather little. They are all smallish (a few centimetres in length), non-colonial and free-swimming. In contrast, there are about 3000 species of urochordates, and these fall into several groups, which are quite distinct from each other. Across the whole range of 3000 species, we find both solitary and colonial forms (Figure 20.1), both mobile and sessile forms, and a 500-fold variation in size, from about a millimetre to half a metre or so.

FIGURE 20.1 Examples of different sea-squirt adults. On the left is a large solitary form, viewed from the side. Water, carrying food material, goes in through the larger of the two holes (the incurrent siphon) and out through the smaller one (the excurrent siphon). On the right is a colonial form, viewed from above, in which the individuals are comparatively small.

One of the main features that unites these two subphyla with our own – the vertebrates – in the phylum Chordata is the possession of a dorsal rod referred to as the notochord at some stage of the life-cycle. We saw earlier that in the vertebrates the notochord is a transient embryonic structure. In the non-vertebrate chordates it is more persistent. In the sea squirts it lasts until the end of the larval stage. In the lancelets it persists throughout life and serves as a kind of spine-substitute in the adult by giving a degree of stiffness – but with some flexibility – to the body and acting as a site for muscle attachment.

It was thought until recently by the vast majority of zoologists that the lancelets are more closely related to 'us' (i.e. the vertebrates) than are the sea squirts and their allies. On the basis of adult body forms, this seemed the most likely pattern of relationship. However, recent comparative molecular studies have suggested that in fact we are more closely related to the sea squirts and other urochordates. Using V for vertebrates, C for cephalochordates and U for urochordates, the old view of the relationships among these three groups can be written thus: ((V + C) + U); the new view is ((V + U) + C). There is not yet a clear consensus on this matter, but the new view is looking increasingly likely to be the correct one.

Chapter 14, which was devoted to examining the world of arthropods, was entitled *Exoskeletons galore*. Although I decided to use a different style of title for the present chapter, it could well have been called *Endoskeletons galore*, because of course vertebrates are characterized by endoskeletons, which may be cartilaginous or bony; and there are more than 50,000 species, each with its own particular version of an endoskeleton. So, to understand the origin of the vertebrates, we need to understand how the endoskeleton arose. But, like the origins of nearly all features that define any animal phylum or subphylum, the origin of the vertebrate endoskeleton is lost in the mists of deep time, to use a slightly modified version of the old saying. As with any major group of animals, it is much harder to understand its origin than its subsequent evolutionary radiation.

The vertebrate endoskeleton did not arise all at once. The skull came first, followed by the vertebral column, followed by jaws and by paired appendages with supporting appendicular skeletal elements – such as the humerus, radius and ulna of the human arm. This sequence of events is implicit in the prevailing current view of the relationships among the living groups of vertebrates, as shown in Figure 20.2. This cladogram underpins the following discussion, and it's a good idea to keep referring back to it as we progress. However, it is also good to bear in mind that any proposed cladogram can be wrong. In the present case, an alternative pattern to that shown in Figure 20.2 is that a common stem leading to lampreys and hagfishes diverged from 'the rest', with these two forms diverging from each other later.

According to the scheme shown in Figure 20.2, the most primitive 'vertebrates' alive today are the hagfish. These are elongate, fish-like forms with few externally visible features. Hagfish even stretch the limits of what we can call a vertebrate. The reason I used quotes around 'vertebrates' above is that hagfish do not have vertebrae. They do, however, have a skull, made out of quasi-cartilaginous material. Although they have a skull, they lack jaws. They also lack the paired fins (pectoral and pelvic) that more advanced fish have. They represent the earliest lineage (at least of those with extant forms) to have diverged from the rest of the vertebrates.

The next lineage to diverge – again sticking with groups that have some extant members – was the one that led to lampreys. These are not unlike hagfish in being elongate and rather featureless, without jaws or paired fins; a lamprey is pictured with a hagfish in Figure 20.3. The main skeletal difference, which of course is not apparent from an external view, is that lampreys have rudimentary vertebrae as well as a skull. Many of today's lampreys are parasitic on fish, using their jawless,

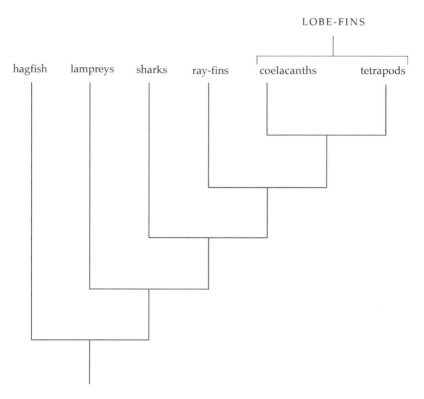

FIGURE 20.2 Cladogram of the vertebrates. Note that the land vertebrates need to be formally considered as 'fish' if we are not to end up with a paraphyletic group. Of course, in more general discussion 'fish' is still restricted to aquatic vertebrates. In the picture, lobe-finned fish are represented by coelacanths, cartilaginous fish by sharks. For illustrations of hagfish and lamprey body forms, see Figure 20.3. Note: there is not a unanimous view that the cladogram shown here is correct with regard to the hagfishes and lampreys (see text).

sucker-like mouth to attach to the fish, rasping at its skin and flesh with their many teeth. However, such an ecological niche would have been impossible when the very first lampreys began to evolve, because there were no true fishes at that point in evolutionary time. What the feeding ecology of the earliest lampreys was is unknown.

The true fishes that I referred to above have jaws and paired fins. There are two main groups of these: the cartilaginous fishes – including sharks and rays – and the bony fishes, the group (called Osteichthyes) into which most of today's fishes fall – for example the familiar trout, salmon, mackerel and haddock. There is fragmentary fossil evidence of cartilaginous fishes from as far back as 450 MYA, with the first

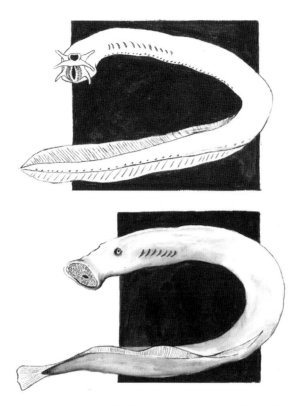

FIGURE 20.3 Two primitive, basally branching vertebrates: the hagfish (top) and the lamprey (bottom). Neither of these have jaws; hence they are sometimes referred to as agnathans (meaning jawless).

near-complete fossil shark dating from considerably later – about 410 MYA. Fossil evidence of bony fishes begins to appear around the same time – the earliest known fossil at the time of writing dates to about 420 MYA.

Although cartilage appeared before bone in the earliest period of vertebrate history, there was not a single origin-of-bone event as there probably was for the origin of jaws. Bone seems to have arisen (and been lost) on several occasions. Also, there are different types of bone, including external bone derived from skin. This was a feature of many armoured extinct forms, such as the fishes called placoderms (plate-skin), which diverged from the rest of the fishes before the cartilaginous–bony fish split. The more general point that *many* features of animals can arise and be lost several times should always be borne in mind when considering the evolutionary history of any animal group, as an antidote to the risk of oversimplifying the actual pattern of events over time.

Just as the cartilaginous and bony fishes first appear at roughly the same time in the fossil record, so also do the two main subgroups of bony fishes. These subgroups are called the ray-finned and the lobe-finned fishes; and understanding the difference between them is crucial for understanding the origin of the land vertebrates – a topic that we will deal with in the next chapter.

The key issue here is the nature of the internal skeletal support for the pectoral and pelvic fins. In fishes of the ray-finned group, these fins are typically thin; inside them are many long thin skeletal elements called rays – think of them as being like the spokes of a wheel, or more specifically like a section or slice of a wheel, with spokes radiating out over a span of perhaps 30–45 degrees (Figure 20.4). Also shown in the figure is the skeletal support of a coelacanth – the not-really-extinct fish that was described in Chapter 15. Here, there are several chunkier bones in the proximal part of the fin. And, in association with this, the fins, when viewed from the outside, look chunkier or lobe-like, as we saw earlier. This is true of all the lobe-finned fishes, not just coelacanths, including several extinct groups and also the lungfish, of which there are extant member-species.

The land vertebrates, or tetrapods, arose from the lobe-finned fishes. Or, from a cladistic perspective, the tetrapods, including ourselves, *are* lobe-finned fishes, because if they are excluded then the lobe-finned fish

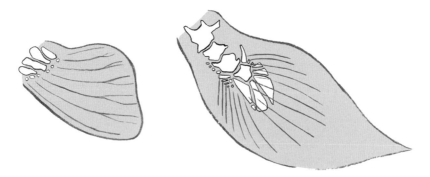

FIGURE 20.4 Skeletal elements in the fins of ray-finned (left) and lobe-finned (right) fishes. Note the chunkier nature of the latter at the proximal end of the fin, the example shown here being the coelacanth.

group is not monophyletic, as can be seen from Figure 20.2. Personally, I think that considering ourselves and other land vertebrates as fish is a step too far in terms of common sense; however, the cladistic logic for it is clear enough. Anyone interested in reading more about our own fishy nature should read the 2008 book *Your Inner Fish*, by the American palaeontologist Neil Shubin. For a broader account of vertebrate evolution than is given either in Shubin's book or in the present one, see *Vertebrate Life*, by Harvey Pough and co-authors, the sixth edition of which was published in 2002.

It's interesting to consider the numbers of extant species belonging to the various groups so far discussed. These can be considered not just as numbers but also as proportions of the overall total of about 50,000 species of vertebrates. We can look at the various groups in their left-to-right order in Figure 20.2.

Taking this approach, and giving very approximate round numbers, there are fewer than 100 species each of hagfish and lampreys. There are about 1000 species of cartilaginous fishes, in contrast to about 30,000 species of bony fishes (excluding tetrapods). Those 30,000 are almost all ray-finned fishes: only about 10 of them are lobe-fins. The land vertebrates number a little over 20,000 species. Thus in terms of the groups shown in Figure 20.2 there is a very uneven distribution. Of course, the relative species diversity of these groups has changed a lot over evolutionary time. To adapt a caveat from the financial world, present performance (in terms of the number of species that a group includes) is no guide to either past or future performance.

Notice that our whirlwind tour of the origin and evolution of vertebrates has so far been based on morphological features of adults. But what about development? All changes in adult forms over evolutionary time take place through modification of the developmental processes through which adult features are made. And what about developmental genes? We have seen (Chapter 19) that there is an 'animal toolkit', with a relatively small number of families of developmental genes, which is highly conserved among different animal groups. But it has not been completely conserved – it too has been modified over evolutionary time. Indeed, if it had not been, then we would not find the great disparity of animal forms that we do.

There is not space to go into all of the developmental and genetic changes underlying the origin and evolution of the vertebrates in an introductory book like this one. But we can examine a few examples, as follows.

First, we need to backtrack to something rather crucial in the evolution of the vertebrates that we have not yet discussed: the origin of the head. If you think about it, phyla of animals without particularly elaborate heads, or indeed without heads at all, vastly outnumber those with significant heads. There are no heads in the sponges or in the cnidarians (jellyfish *et al.*). There are only minimal heads in those many worm phyla that we discussed in Chapter 10. The echinoderms (starfish, sea urchins and their allies) have lost their heads, so to speak. The basal non-vertebrate chordates (lancelets and sea-squirt tadpoles) have little in the way of a head. Most molluscs have a small head (snails), or essentially no head (bivalves): only members of the relatively small group called the Cephalopoda (including octopuses) have a head that has been highly evolutionarily elaborated. The only phylum apart from the vertebrates where having a significant head is the rule rather than the exception is the Arthropoda. So the evolution of an elaborate head is interesting for its rarity as well as for its potential in terms of behaviour and, in some cases, intelligence.

Interestingly, in vertebrate embryogenesis many head structures, including cranio-facial bone, derive from neural crest cells – a population of migratory cells that originates during neurulation in the dorsal midline area of the embryo (refer back to Figure 16.3). This cell-type is only found in vertebrates. The headless lancelets have no such cells. Neither do the sea squirts, though they have pigment-producing cells that may be evolutionarily related. So the neural crest cells of vertebrates are an evolutionary novelty (inasmuch as anything can be really novel in evolution: see Chapter 23), possibly having evolved from a pigment-producing cell-type in the last common ancestor of vertebrates and urochordates.

And now from the origin of heads to the origin of a particular head structure: jaws. It seems as certain as evolutionary hypotheses can be that jaws evolved from gills. To be more specific, the jaws of early fishes evolved from the anterior gill supports of jawless ancestors that probably resembled today's hagfish and lampreys. Although I haven't discussed gills up to now, it is well known that aquatic vertebrates almost all use gills to extract oxygen from water (whales are an exception, of course), just as terrestrial vertebrates almost all use lungs for extracting oxygen from air. In aquatic vertebrates, gills typically occur in pairs, opening on the left and right sides of the anterior end of the animal, in what is called the pharyngeal region. The number of pairs of gills is quite variable among – and sometimes within – different groups. The overall range is from three pairs to 'many' – for example about 15 in some hagfishes, though there is a complication that in these animals the number of pairs of gills and of gill openings is not exactly the same.

The jaws of vertebrates that possess them have a similar embryo-logical origin to the anteriormost pair of gills in lampreys; or, to be more specific, they have a similar origin to the gill *arches*. Gills are normally supported by cartilaginous or bony rods called either gill arches or branchial arches. A shared developmental origin is suggestive that the jaws evolved from the first pair of gill arches of an ancient, lamprey-like ancestor. But how would the intermediate stages of this transition have been beneficial? One hypothesis is that the first stage in the enlargement and shape-change of the anteriormost pair of gill arches was selectively advantageous because it produced a structure that enabled a stronger pumping of water over the gills. If this is so, then the evolution of jaws began as an improvement in gill function, and only later did further modification take place for reasons related to feeding rather than breathing.

Regarding evolutionary changes in developmental genes associated with the origin and early evolution of the vertebrates, an interesting example to look at is the Hox gene family. Recall, from the previous chapter, that this family is found in all bilaterian animals, and its member-genes have a major role in patterning the antero-posterior, or head-to-tail, body axis. In the fruit-fly *Drosophila*, where studies on these genes started, there are eight of them, and they are found in two complexes close together on a particular chromosome. This arrange-ment is called a split cluster; most insects other than flies have a single (i.e. non-split) cluster with a similar number of genes.

The exact number of Hox genes varies among invertebrates, but is usually in single figures. It's also a single figure in one of the two non-vertebrate chordate groups (urochordates: 9); but is higher in the other such group (cephalochordates: 14). In the vertebrates the number is much higher, because of whole-genome duplications. The number of clusters can be up to eight and the total number of Hox genes can be anything up to about 100 (because each cluster can have up to the full complement of 14 genes, though usually the number is less because of gene losses).

Increases in Hox gene numbers in the vertebrates have been implicated in the evolutionary increases in complexity of body form that have taken place from the proto-chordate groups to the jawless fishes such as the lamprey to the 'higher' vertebrates. However, this idea needs some dissection, as follows.

Comparisons among various vertebrate genomes show that there were two key duplication events. The first of these occurred in the stem species of the jawed fishes and the second in an important stem lineage within the ray-finned fishes, one that led to the bulk of today's fishes – the group called teleosts. The first of these tree positions connects well with the increases in complexity that took place in the group of vertebrates that the jawless lamprey lineage diverged away from: the origin of jaws and of paired fins. Although it has been argued that the second duplication was connected with the early evolutionary radiation of teleost fishes, this radiation represents an increase in diversity more than in complexity. There was no similar scale duplication in the stem lineage leading to the land vertebrates (the tetrapods), which most biologists would argue are more complex than teleost fishes. Also, we know that further duplication events have occurred in subgroups of ray-finned fishes that exhibit no evidence of exceptional diversity or of increases in complexity.

One way to interpret the overall pattern is that some evolutionary increases in complexity are caused – or at least facilitated – by genome duplications while other increases in complexity manage to happen without them. This makes sense, because it is hard to see how duplication *per se* can cause an increase in the complexity of the body forms of the animals concerned. Rather, it is events occurring post-duplication that really matter. In particular, the divergence in function of the duplicated copies of particular genes is what counts. This can take place after the duplication of single genes as well as after whole-genome duplications. If the single genes are key developmental ones, then their duplication and divergence may be sufficient to fuel a complexity increase.

The divergence in function that occurs in duplicate copies of developmental genes may involve two things. First, there may be changes in the gene product – that is, the protein the gene concerned makes. Second, there may be changes in the way the gene is regulated by those other genes that are developmentally upstream of it, so that it comes to be switched on in different places and/or during different periods of developmental time than previously.

One example of a new role of developmental genes in the jawed fish is a role of specific Hox genes, such as *hoxD13*, in the development of paired fins. (The name of this gene refers to it being in position 13 in Hox complex D.) This gene's product seems to specify which region of the developing fin-bud is posterior. Such a role would not be possible in the lamprey, which does not have paired fins. Interestingly, *hoxD13* also has a role in specifying the posterior region of the limb-bud in the embryos of the land vertebrates – i.e. the tetrapods. But how did the tetrapods themselves arise? In other words, how did vertebrates invade the land? And when did they do so? These are the questions to which we turn in the next chapter.

21 From water to land to water

Unlike arthropods, several groups of which invaded the land independently, vertebrates invaded the land just once in the course of evolution. Thus the group of land vertebrates – the tetrapods – is a monophyletic one. Much research has gone into the question of how, and when, this invasion took place, and as a result we now have a reasonably good idea of the answers to both the 'how?' and the 'when?' questions. One of the main features that changed in the vertebrate animals that invaded the land was the nature of their appendages: from the fins that help to power fishes through water to the legs that we and most of our fellow tetrapods use to walk on solid ground. So the invasion of the land by vertebrates is sometimes equated with the fin-to-limb transition – though strictly speaking these two things are not exactly the same, and they may not have occurred at exactly the same time as each other, as we'll see.

Another of the main changes that occurred in the vertebrates that invaded the land was a shift from breathing in water to breathing in air. So we could also refer to the gill-to-lung transition, though this phrase is less often used. There is an interesting difference between the two transitions, which becomes apparent when terrestrial vertebrates go back to the water, as in the case of whales, in relation to their reversibility. Fins can be re-evolved quite easily, it seems. In contrast, it appears to be much more difficult, perhaps even impossible, to re-evolve gills.

Let's start with the 'when?' question. As far as we know, there were no terrestrial vertebrates 400 million years ago (MYA). But by 340 MYA there were many. The invasion of the land probably started around, or somewhat before, 360 MYA. The fossils of several transitional forms are known from this period. As usual, hard parts fossilize best, so the story can be told largely in terms of fossilized skeletons – or bits of them. We already saw (in the last chapter) that tetrapods evolved from lobe-finned fishes; our ancient aquatic ancestors were somewhat similar to today's coelacanths, whose fin supports were pictured earlier (Figure 20.4). Since such fins already had stout bony supporting rods, the next steps in evolving a leg-and-foot from a lobe-fin were further strengthening, some rearrangement of the bony elements, and a separation of the surrounding flesh into distinct digits.

When I was a student studying evolution in the 1970s, I learned a lot about the evolution of the *pentadactyl limb*. This phrase was (and sometimes still is) used to refer to limbs that end with five toes (or fingers). The pentadactyl arrangement is typical of the primates, including, of course, ourselves. It is also typical of several other mammal groups. Those that differ from the five-toed pattern can usually be thought of as having started with it but having then been modified, over the course of evolutionary time, so that some digits were reduced. This is the case, for example, with mammals that have hooves. There are two main groups of these – referred to as odd-toed and even-toed.

The rhinoceros and the horse are members of the odd-toed group. The horse's hoof is a large digit 3, with the other digits having been lost. The rhino's hoof is centred on digit 3, but has flanking digits 2 and 4 (Figure 21.1). These numbers may make more sense if we relate them

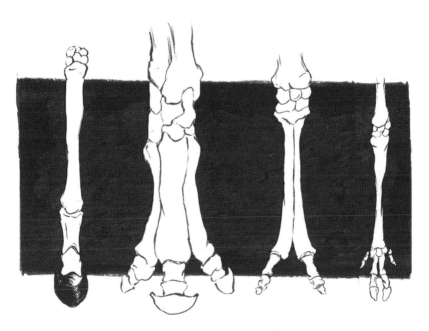

FIGURE 21.1 The main bones in the feet of a variety of hoofed mammals. On the left are two representatives of the odd-toed arrangement in perissodactyls (left – horse, one toe; right – rhino, three toes). On the right are two representatives of the even-toed arrangement in artiodactyls (left – camel, two toes; right – deer, four toes, though only two are normally used for weight bearing).

to the human hand, in which the thumb is digit 1, the index finger digit 2, the middle finger digit 3, the ring finger digit 4, and the little finger digit 5. The deer and camel are members of the even-toed group. A camel's hoof is based on digits 3 and 4, while a deer's hoof also has these two digits as the main load-bearing ones, but they are flanked by smaller digits 2 and 5 (Figure 21.1).

Of course, comparisons with the human hand are made because of the familiarity of our own digits and not because primates were ancestral to mammals with hooves. However, comparison of mammals like the horse and the camel with a more general five-toed pattern is made because it is thought that these hoofed creatures evolved from a five-toed ancestor which, like us, had five digits per foot. That is, the mammalian limb, and indeed the tetrapod limb more generally, was thought of as starting with five toes per foot: this was considered to be the ancestral pattern.

However, discoveries of fossils from the 390–360 MYA period have shown that this idea is probably wrong. The earliest fossil tetrapods had more than five digits per foot – some had six, some had seven and some had eight. So it seems that an early period of instability in digit number came first. This was then followed by stabilization on the five-digit pattern; after that, there were further reductions in some lineages, such as the hoofed mammals discussed above together with many others.

There are quite a few famous fossils that illustrate the fin-to-limb transition. However, so that this does not become an exercise in learning names, I'm going to focus on just one of them: the early tetrapod called *Acanthostega*. Fragments of fossilized skeleton of this animal were first found around 1930. However, it was not until the 1980s that the first complete limb was found. It was described by Mike Coates and Jenny Clack in a paper in the journal *Nature* in 1990. (Both Coates and Clack had been research students of Alec Panchen.) At this stage it had become apparent that *Acanthostega* had eight digits per foot (Figure 21.2), a finding that was popularized by Stephen Jay Gould in one of his books of collected essays, which he entitled *Eight Little Piggies*. This was published in 1993, and one of the essays it contains tells the story of early tetrapod fossils and their unexpectedly high digit numbers.

Sometime around 370–360 MYA *Acanthostega* and its relatives were flourishing. But in what type of habitat? Bearing in mind that they can be thought of as the first amphibians, did they primarily inhabit water or

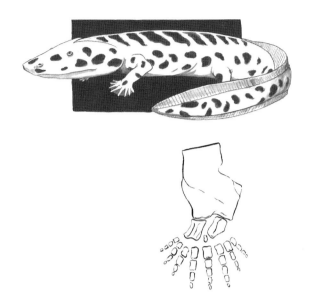

FIGURE 21.2 The early tetrapod *Acanthostega*, which lived about 370–360 MYA. Top: an artist's impression of the whole animal. Bottom: the eight digits, and other bones, of the forelimb.

land? Studies on the bone structure of the forelimbs suggest that they may have been too weak for a truly terrestrial existence.

One hypothesis is that they lived in shallow bodies of water with dense plant material and that their feet evolved to paddle through such thickly vegetated water bodies. If so, then the first stages of the transition from fin to limb occurred for reasons connected with movement through certain types of aquatic habitat; and only later stages of the transition took place for reasons connected with walking on dry land. There is a link, here, with the evolution of jaws from gill arches (previous chapter), where the first steps may have been to do with better breathing and only the later ones to do with better feeding. In general, when a structure evolves in this way, with adaptation first for one reason and then for another, the structure is said to have been exapted; and the overall evolutionary process is called *exaptation*. This process may be of considerable importance in evolution.

I said earlier that the story of the invasion of terrestrial habitats by the vertebrates was dominated by two structural transitions: fins-to-limbs and gills-to-lungs. While these two transitions were indeed keys to the invasion of the land, other changes were involved too. One of the most important of these in allowing the connection with aquatic habitats to be finally broken was the evolution of a new kind of egg.

The eggs of early tetrapods such as *Acanthostega* were probably rather like those of today's amphibians – for example frogs. Most species of frogs lay their eggs (frogspawn) in ponds or other water bodies. So although the adult frogs spend much time on land, eggs and other developmental stages are confined to water. Frogspawn laid on dry land would desiccate and the embryos within it would die. The reason that this does not happen to birds' eggs is that they are much better protected from desiccation.

Birds' eggs are right at the other end of the spectrum from frogs' eggs regarding the degree of protection of the embryo within them. This is why a bird can get away with laying just a few eggs, while a frog has to lay many: the chances of an individual bird's egg surviving are much higher than those of an individual frog's egg. Even when the latter is not dried out, its chances of survival are low – for example because of the great risk of predation. But evolution of the tetrapods produced a better-protected egg than the amphibian one long before the origin of birds. In fact, an early split in the tetrapods produced two groups: the Amphibia and the Amniota, with the latter being named after its evolutionarily improved egg: the amniotic egg that is found in reptiles, mammals and birds – the groups that make up the Amniota.

Many readers will have heard of the medical procedure called amniocentesis. This involves taking a sample of the amniotic fluid that surrounds the human embryo within the embryonic sac – a sort of bag made of amniotic membrane. This is one of several types of membrane in the amniotic egg (Figure 21.3). There is also an outer membrane called the chorion, for example. And in birds, of course, there is also, outside the chorion, a shell. In contrast, an amphibian egg has just a single membrane.

Although the evolution of the amniotic egg allowed a degree of independence from water far beyond that of the amphibians, tetrapod groups vary enormously in the nature of their eggs – bird and mammal eggs could hardly be more different from each other. Tetrapods also vary enormously in the degree to which they are truly terrestrial at various stages of the life-cycle. Many tetrapods have become wholly or partly aquatic again. These include turtles, crocodiles, some snakes, penguins, whales, seals and sea cows.

amnion

chorion

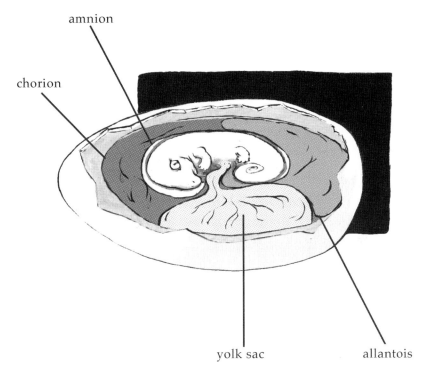

yolk sac   allantois

FIGURE 21.3 The amniote egg, with its various membranes, in particular the amnion, chorion, allantois and yolk sac. The evolutionary invention of this egg enabled those tetrapods possessing it to be more independent of the water than their ancestors (and also than today's amphibians).

Before looking at these back-to-water tetrapods, it's worth discussing the incredible radiation of tetrapod groups on land – and indeed in the air too. Numbers of species give some idea of success, though with the usual proviso that present number of species within any group is no guide to past or future numbers.

Of the approximately twenty-something thousand species of tetrapod, the vast majority remain on land. Some of the biggest success stories are the squamates (lizards and snakes) with about 8000 species, the birds with about 10,000 species, and the mammals with about 5000 species. In each case, a subtraction needs to be made to remove the secondarily aquatic forms (for example, water-snakes, penguins and whales); but again, in each case, these aquatic forms are small minorities.

Although the birds are by far the biggest group of tetrapods to have taken to the air, there are more than 1000 species of bats. And, in

Jurassic times when dinosaurs were the dominant land vertebrates, there were also the flying pterosaurs. It's always harder to estimate the number of species of an extinct group than an extant one. However, many different families of pterosaurs are recognized, and at their peak the number of species was probably well into the hundreds – though I must admit that this is just a guesstimate.

We'll now look at some of the groups of tetrapods – and more specifically of amniotes – that returned to the water. Their evolution tells an interesting story about the potential reversibility of various kinds of evolutionary change, as noted earlier. We'll focus largely on the mammals. Setting aside semi-aquatic mammals such as the otter, there are three groups of mammals that have either largely or completely returned to the water. These are the cetaceans (whales and dolphins), the pinnipeds (fin-feet: seals and walruses) and the sirenians or sea cows (dugong and manatee: Figure 21.4).

Each of these constitutes a monophyletic group that evolved from a single returning-to-water ancestral stem. The closest living relatives to the whales are to be found among the group of mammals with even-toed hooves, such as the hippopotamus. In contrast, the closest living relatives of the seals are probably the bears; and the closest living relatives of the sirenians may be elephants. So it's not a case of a single re-invasion of the water and then a diversification into cetaceans, pinnipeds and sirenians.

Given that fact, it's interesting to look at features of each of the three aquatic mammal clades and see to what extent they are similar or

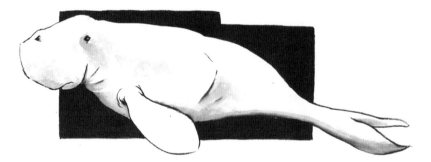

FIGURE 21.4 A dugong or sea cow – a member of the mammalian order Sirenia. Dugongs are marine herbivores. Adults are typically between 2 and 3 metres long. The manatees, of which there are several species, are also included in the Sirenia.

different. Any similarities must be due to convergent evolution, given that the re-invasions of the water were independent. We'll concentrate on the two features already mentioned: the nature of the fins/limbs and the method of breathing. However, it makes sense also to look at overall body form.

It's perhaps obvious, but nevertheless worth stating, that the body forms of all these aquatic mammals are fish-like, or fusiform, to various degrees. That's more true of a striped dolphin than a humpback whale, and more true of a leopard seal than a walrus; but, despite such variation, all the body forms concerned are more streamlined than those of their closest non-aquatic ancestors (hippos, bears and elephants). So it seems that natural selection has no difficulty moulding overall body shape in aquatic mammals, just as it has no difficulty moulding body shape in animals generally.

The limbs of each group have evolutionarily reverted to being fin-like, or in some cases have disappeared or become vestigial. Of the three groups, cetacean and sirenian fin (and tail) arrangements are more similar to each other than either is to the pinnipeds. Whales and dolphins typically have a pair of pectoral fins, lack a pair of pelvic fins, and have a tail, or fluke, that is flattened in the dorso-ventral axis rather than the left-right one, as in fish. Given their separate origins, these convergent features are impressive. Pinnipeds are similar in having pectoral fins (or flippers). However, they differ from cetaceans and sirenians in the anatomy of their rear appendages. Instead of a tail, they have a pair of pelvic flippers – modified hind-limbs. The term 'flippers' is used because the transformation from the separate digits that characterize land mammals to the fused digits of a mammalian fin is less complete than in the other two groups. So there is still convergence when comparing, say, a seal with a whale, but it is less pronounced than when comparing a dugong with a whale.

Now we turn to the question of how animals of these three groups breathe. The air-breathing of whales (and dolphins) is familiar to almost everyone because of the exhalation of waste gases through their blow-holes, resulting in a spectacular upward blasting of the surrounding water; this is often shown on television natural history programmes. What is less well-known is that the blowholes are relocated nostrils. Some whales have two blowholes, corresponding to the original two nostrils of their terrestrial ancestors. Others have just one (the second nostril having evolved into an echolocation system). This evolutionary repositioning is brought about via an embryonic shifting of the nostrils from the front of the head to the back, so this is an example of recapitulation – in the more enlightened sense of that word, as discussed in Chapter 17. Whale and dolphin blowholes lead via a trachea to the lungs. So the arrangement is rather similar to that of land mammals, except that the link between the oesophagus and the trachea has been broken, with the result that whales can't choke on their food.

Pinnipeds and sirenians also breathe air using lungs, but they have retained a more anterior position of the nostrils than in whales. So, with respect to breathing mechanisms, these two groups are more similar to each other than either is to the cetaceans. However, this cannot be regarded as a case of convergent evolution between pinnipeds and sirenians, because nostrils are ancestral whereas blowholes are derived.

So far, we have looked at the features that have evolved in aquatic mammals; now it's time to look at features that we might have expected to have evolved but which have not. This approach might be called looking at possible animals as well as actual ones. It's an approach that can be highly informative, as noted in Chapter 1 in the context of the lack of evolution of dragons, but is not used nearly enough. Comparison of the actual and the possible reveals not just what evolution can do but also what it seems not to be able to do.

The whale's blowhole has often been seen as a wonderful adaptation. But wait a minute. It works, but surely something else – gills, to be specific – would be better? A whale needing to come up to the surface to breathe air through its lungs is a bit like a human who had to jump into a swimming pool from time to time in order to breathe water through gills. So it seems reasonable to ask: given that aquatic mammals have been able to re-evolve fins, why have they not also been able to re-evolve gills?

This question is given even more weight if we look at non-mammalian groups of amniotes that have returned to water. Penguins have impressive flippers; but they still breathe air using lungs. Maybe that's not so surprising, because penguins spend some time on land, which whales and dolphins do not. However, turtles also have flippers but lack gills and so have to breathe air. And there's another amniote group that re-invaded aquatic habitats in evolution and ended up with flippers but no gills. It's less familiar to us than penguins, turtles, whales

FIGURE 21.5 A plesiosaur – an extinct marine reptile. These were common ocean animals in the Jurassic period (from about 200 to 145 MYA). The ancestors of plesiosaurs were terrestrial, so plesiosaurs represent a secondary return to an aquatic existence, just as do whales.

and so on simply because all its members are extinct. The animals I'm referring to are the plesiosaurs – aquatic reptiles that were common in the oceans of the Jurassic period (Figure 21.5).

Why has evolution not been able to re-invent gills for breathing water in whales, dolphins, seals, sirenians or plesiosaurs? The answer hardly lies in the fact that being able to breathe water would not be an advantage. If that's true, then the answer must lie instead in the availability of some types of variation for natural selection to act on, and the unavailability of other sorts of variation. So you can see where this line of reasoning is taking us: to the nature of variation itself.

22 Variation and inheritance

The existence of variation is a prerequisite for natural selection to be able to work. Furthermore, the variation must be at least partly heritable, because if it is not then any differences between variants in their survival rate or their number of offspring will not be carried through to the next generation and subsequent ones. Darwin made the importance of variation very clear in the way he structured *On the Origin of Species*: the first chapter is called "Variation under domestication" and the second "Variation under nature". Thus, to get to his main contribution to evolutionary theory, he took a route that started with variation. ("Natural selection" is his fourth chapter).

Not only must variation be heritable, but the mechanism of inheritance must work in such a way as to maintain it. This may seem a strange point to us now, in the twenty-first century, but Darwin was acutely aware of the problem that if inheritance worked by some kind of blending of the contributions of the two parents, then in just a few generations whatever variation there was would have been eliminated. And at the time when Darwin was writing his *magnum opus*, the idea of blending inheritance was commonplace; it was not until some years later, in 1866, that a different mechanism of inheritance was proposed by the Austrian monk Gregor Mendel. Even then, this different mechanism, which is referred to as particulate rather than blending inheritance, languished in obscurity until its rediscovery in 1900.

The way in which blending inheritance is problematic for natural selection can be seen using the following example. Imagine a kind of mammal in which males and females have identical patterns of variation in body size (unlike humans) and whose body form is such (say, shrew-like) that the most obvious measure of body size is length, as opposed to height. Suppose that the overall range of variation in adult body length in a population of these creatures is from 6 to 10 centimetres. With blending inheritance, a mating between 6 cm and 8 cm parents will produce 7 cm offspring. Similarly, a mating between 8 cm and 10 cm parents will produce 9 cm offspring. If one of the offspring from the first mating then mates with one of the offspring from the second, all of their offspring will be 8 cm. So, it's easy to see how, in a system of blending inheritance, whatever variation was present initially will be 'blended out of existence' within a few generations, leaving a homogeneous population upon which natural selection cannot act.

The original mammal may have been somewhat shrew-like; it was almost certainly small. Yet from it natural selection has produced immense differences in size between species, so that we now have mammals all the way up to the size of elephants and whales. This result has only been possible because inheritance works in a particulate rather than a blending manner.

The term *particulate* refers to the fact that each parent involved in a mating contributes a 'particle' to the offspring. The two particles together determine the body length (or other character) of the offspring. However, when those offspring become mature and embark on their own reproductive journeys, the particle that came from the mother and the corresponding one from the father separate again and go into different eggs (in a daughter) or different sperm (in a son). So their association was only temporary. In this form of inheritance, variation can be maintained indefinitely in the population.

There is, however, another potential problem, which led to a strange controversy among biologists in the early 1900s. If inheritance is particulate, we might expect to find that the individuals in a population vary in a discrete way. Indeed, anyone who is familiar with Mendelian genetics will be aware of the numerical ratios between different discrete types of offspring that are to be expected in certain types of mating. For example, when Mendel crossed tall and short pea plants with each other, he got an approximately three-to-one ratio of tall to short offspring two generations later, with no intermediate-height plants.

This discrete variation is very different from the continuous variation in characters such as body size that we see in nature. For example, in humans, both adult male and adult female heights have *distributions* of values. These can be represented not as ratios between two discrete types ('tall' and 'short'), but rather as bell-shaped (normal) curves, each symmetric about its average, with all possible variants between the upper and lower extremes being represented (Figure 22.1).

Shortly after the rediscovery of Mendel's work in 1900, some biologists thought that the continuous variation in characters such as human height, or body size in animals more generally, was not inherited

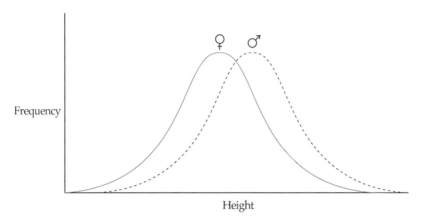

FIGURE 22.1 Two normal distribution curves: one for the variation of adult female height, the other for the variation in adult male height, in some particular population of humans – in some particular geographic area. Note that the curves are similar in shape (they are bell-shaped or normal) but that each is displaced relative to the other.

because it did not behave in a Mendelian way, with recognizable ratios between offspring that had discretely different character values. So they wrote off this variation as being evolutionarily irrelevant. However, in 1918 the British geneticist and statistician R. A. Fisher showed that if lots of different 'particles' contributed to the value of a single character such as body length, Mendelian ratios would turn into continuous distributions. We now, of course, refer to these particles as *genes* – Mendel had called them factors. In addition, we have known, since the 1950s, the molecular nature of genes. Thus we can deal with them at the DNA level as well as the level of the organism and its character-value – which is often called its *phenotype,* though I'm trying to avoid this word where possible for reasons that will become clear shortly.

To summarize: the combination of the work of Gregor Mendel in the 1860s, R. A. Fisher in the early 1900s, and finally Watson and Crick in 1953, has taken us from a state of ignorance about the mechanism of inheritance to a state of reasonably thorough understanding. And interestingly, this has been a series of advances in which each has in some way confirmed or added to the ones that went before. So it has not been a case of overturning an old incorrect hypothesis and replacing it with a new one – apart from at the very beginning of this series of advances, when the incorrect hypothesis of blending inheritance was disposed of. Fisher showed that Mendel's particulate system was right, not just for cases of discrete variation but for cases of continuous variation too. Later, Watson and Crick showed us the physical nature of Mendel's factors.

But that is far from the end of the story. There have been further advances in the study of variation and inheritance over the last 60 or so years. We'll look at these shortly. And there are almost certainly major advances in our understanding of variation yet to come. To see why I believe this, let's dissect the concern I expressed above at the use of the word phenotype.

As you may already know, this is one of a pair of terms – the other being *genotype*, a word I have no problem with at all. Conventional usage is that genotype refers to the combination of two copies of a particular gene (which does a particular job) that an animal has, one copy inherited from its father and one from its mother. Suppose that, among the many genes that contribute to the body length of the hypo-thetical mammal introduced earlier, there is one – call it *blg14* – that resides at a particular point on a particular chromosome within the nucleus of every cell in the animal's body. The rationale for this code may be apparent; in case not, it stands for 'body length gene number 14' (out of some larger total number of such genes – say 30). Suppose further that there are two different versions of this gene in the population, one conferring increased body length, *blg14-i*, the other conferring decreased length, *blg14-d*. Each individual animal in the population must have one of three genotypes – it may have two copies of *blg14-i*, two of *blg14-d*, or one of each (the same-copy types being called *homozygotes*, the different-copies type being the *heterozygote*).

The reason I have no problem with the use of *genotype* is that it really does refer to a *type*: it doesn't change over the animal's development or, more generally, over its entire life-history. It's fixed. But *phenotype* is not a type at all. It is often used to refer to the adult, whereas most characters of an animal, including its body length, are things that change during development. Typically, a character such as body length has a trajectory over developmental time. In mammals, which have direct development and no larval stage, the trajectory for body length generally goes upwards, but at different rates during different stages. Two adult animals in the same population that have different body lengths may end up that way because their developmental trajectories diverged at an early embryonic stage such as neurulation, a later one such as organogenesis, or sometime during the lengthy period of post-embryonic growth.

The widespread use of *phenotype* can be considered as symptomatic of the non-developmental thinking that pervaded much (but not all) of evolutionary theory in the mid-twentieth century, before the advent of evo-devo in the 1980s. Such thinking was perhaps excusable then but it definitely isn't now. Since evolution works by modifying the developmental trajectory, it makes no sense to ignore this trajectory and to think only of its end-product. So evolutionary thinking needs to be re-cast in explicitly developmental terms. This is an area in which I think there will be considerable advances over the next few decades.

But before speculating about what these advances in the future may be, let's return to the present and the recent past. I said earlier that there have been advances in our understanding of variation and inheritance since the work of Watson and Crick. We should examine what these are. I'm going to divide them into (a) advances in understanding the variation in a single character, such as body length, and (b) advances in understanding the co-variation of two or more characters. We'll take them in that order.

Following Fisher's demonstration that continuous variation could be produced by the combined action of lots of genes all having a small effect on the character concerned, there was a period of a few decades in which the general view was that 'lots' meant about 100 or so – at least to an approximation. There was also a general view that each of these 100 or so genes had a very small, perhaps individually undetectable, effect on the character value.

However, roughly coincident with the birth of evo-devo in the early 1980s, quantitative geneticists began to analyse the genetic basis of characters that had continuous variation. These analyses fall under the heading of QTL studies; so let's look at what these initials mean. The Q stands for *quantitative*; in the present context, this is just another way of saying continuous. The T stands for *trait*, which is another way of saying character – so, for example, body length can be described either as a character or as a trait. The L stands for *locus*, which is effectively another word for gene, though I should expand on that rather brief statement. I talked, earlier, of a gene as being a stretch of DNA that 'does a particular job'; and I indicated that there could be different versions of this gene that did the job a bit differently. What I did not emphasize earlier was that each gene has a very specific location on a chromosome. The L for locus is referring to a gene by this location (and the different versions of the gene that I mentioned are referred to by geneticists as *alleles*). So you can talk about a gene and its different versions or a locus and its different alleles; the former is a more casual version of the latter.

So now it's clear what QTL means. And, although I referred above to QTL studies, another commonly used phrase is QTL mapping. This is because the studies concerned try to determine three things: first, the number of genes contributing to a quantitative character; second, their magnitudes of effect on the character; and third, the map position of each gene in the genome – on which chromosome it is found, and approximately where along the length of that chromosome.

This is not the place for an explanation of the methods used in these studies, which are complex. So we'll skip over those and look at the results. The main conclusion to emerge has been that a typical quantitative character in a typical population does not have a genetic basis that's composed of about 100 genes all of tiny and equal effect, but rather a genetic basis of somewhat fewer genes – hard to specify, but often in the range from about 20 to about 50 – with quite variable effects. That is, some of the genes contribute much more than others to the variation. It's not a case, though, of there being different classes of genes – say those with big, middling and small effects. There's a complete spread from very small to quite large. With regard to the map positions, typically the genes underlying the variation in a quantitative character are scattered all over the place, across many chromosomes.

In some cases individual QT loci can be identified. For example, the size and shape of a bird's beak (Figure 22.2) is influenced by, among many others, the genes that make the proteins called Calmodulin (a contraction of calcium-modulated protein) and BMP4 (bone morphogenetic protein 4). Again, this is not the place for detailed description of the methods involved. The important thing to note is that these two proteins were already known to play important developmental roles in animals. So the genes that make them would be two of the usual suspects for being among the QT loci for a character – adult beak size/shape – that is the outcome of a developmental trajectory. On a more general note, it's worth emphasizing that the studies discussed above fall into a fertile area of research where quantitative genetics meets evo-devo.

Now we need to broaden out from our single-character starting point and think about the way in which, and the extent to which, two (or

FIGURE 22.2 The beaks of two species of Darwin's finches. The species shown both belong to the group known as ground finches. *Geospiza magnirostris* has a much larger and chunkier beak, and indeed is a larger bird overall, than its close relative *Geospiza fuliginosa*.

more) characters co-vary. This is an issue on which Darwin and Wallace differed. However, despite the fact that much has been written on the life and work of both men (especially the former: the production of publications about him is sometimes referred to as the Darwin industry), it is not a difference between them that gets much attention. A single quote from each of the two founders of the theory of natural selection will suffice to indicate the difference to which I'm referring.

In chapter 5 of *On the Origin of Species*, Darwin includes a section entitled "Correlation of growth". He explains this phrase as follows:

I mean by this expression that the whole organisation is so tied together during its growth and development, that when slight variations in any one part occur, and are accumulated through natural selection, other parts become modified. This is a very important subject, most imperfectly understood.

In contrast, in chapter 3 of his 1889 book, *Darwinism*, Wallace stresses the importance of variation in general and then goes on to state:

> Yet more important is the fact that each part or organ varies to a considerable extent independently of other parts.

So we have a situation in which both of the great men emphasized the importance of variation. But when it came to the issue of the extent of co-variation versus independent variation of different characters, Darwin emphasized the former, whereas Wallace stressed the latter. Is this a case of one man being right, the other wrong? Or are they simply pointing to opposite sides of the same coin in that there is both a degree of co-variation and a degree of independent variation when we look jointly at two (or more) characters? The latter view is probably correct, but that does not necessarily remove any possible disagreement. Rather, the argument becomes more complex. Are the two sides of the coin of equal weight? Or is this one of those notorious unbalanced coins, purportedly used by gamblers but rarely if ever encountered? It may be that co-variation is common, independent variation rare. Or the opposite may be true. The two scenarios have important consequences for the ways in which natural selection can, and cannot, bring about evolutionary modification of animals.

One case in which there is usually much co-variation involves the lengths of the forelimbs and the hindlimbs of tetrapods. This applies both within a species (usually, a human with longer legs than another will also have longer arms) and also between species (foxes have longer forelimbs than mice; they also have longer hindlimbs). However, natural selection can sometimes act on correlated characters in such a way that it breaks the co-variation. A good example of this is the evolution of tyrannosaurs, with their massive hindlimbs and their tiny forelimbs – presumably from an ancestor that had its fore- and hindlimbs more similar to each other in length.

Two recent sets of experiments relating to this issue were conducted by the English biologist Paul Brakefield, the Portuguese biologist Patricia Beldade and their colleagues. They were working on butterflies, and in particular on a species called the African brown. Although indeed mostly brown, as its name would suggest, individuals of this species have eyespots on their wings that are composed of a series of concentric circles of different colours. Performing experiments in the laboratory

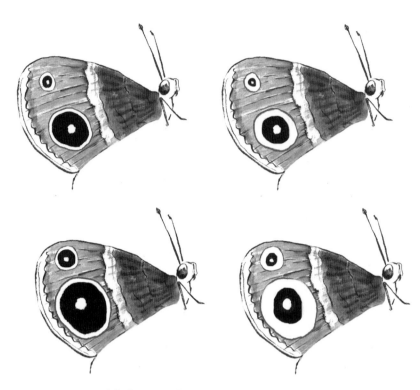

FIGURE 22.3 Simplified picture of variation in eyespot size and pigmentation in the butterfly known as the African brown. In reality most individuals of this species have several pairs of eyespots. Here just two, on the ventral forewing, are shown – enough to illustrate the co-variation that exists in natural populations (a) in their diameters (compare top with bottom) and (b) in their pigmentation patterns (compare side-to-side). The outer grey circles here are in reality a goldish colour.

designed to mimic the effects of natural selection in the wild, these biologists investigated whether selection could overcome two kinds of co-variation found in natural populations of this species.

To simplify things, we'll consider the situation in which butterflies have just two pairs of eyespots (Figure 22.3; in the wild, they usually have several pairs). The first experiment performed by the research group of Brakefield and Beldade involved trying to select for bigger 'eyespot 1' and smaller 'eyespot 2', in other words attempting to break the natural co-variation in size between the two spots. This attempt was successful; the paper describing it was published by Beldade *et al.* in 2002. The second experiment attempted to break the natural co-variation in the pigmentation of the two eyespots, specifically selecting for more black pigment (and less gold) in 'eyespot 1' and the opposite combination in 'eyespot 2'. This attempt at breaking the natural co-variation failed: see the 2008 paper by Allen *et al.*

What the results of this pair of experiments tell us is that the degree to which natural selection can break co-variation and turn it into independent variation differs depending on which pair of characters is involved. It tells us nothing directly, though, about the relative commonness of co-variation and independent variation, because both experiments involved pairs of characters that *did* co-vary. I suspect that the relative commonness of the two types of variation would again depend a lot on the choice of characters. But we are a long way from being able to replace that rather vague statement with something more quantified. Thus it is still the case, as Darwin said, that this is "a very important subject, most imperfectly understood".

All of this chapter has involved the sort of variation that exists within an animal population at a given moment in time. This is called the *standing variation*, and it is the raw material that natural selection can work on, at that moment in time, to produce evolutionary modification. Since particulate inheritance acts to maintain the standing variation, the raw material available for natural selection will usually not be much different from one generation to the next. However, there must inevitably be limits to what can be achieved by selection on the basis of the standing variation. It may be the case that, sometimes, evolution can only proceed so far until one or more unusual mutations, of a kind that are not normally present in the standing variation, occur. In particular, this may be the case in the origin of evolutionary novelties, which are the subject of the next chapter.

23 Evolutionary novelties

Evolutionary novelties represent a topic of immense current interest, and yet they are notoriously hard to define. It's probably best to explain what they are by giving examples, and contrasting these with examples of what can be called routine evolutionary change.

In fact, we have already come across two examples of evolutionary novelties at the level of the cell. These were the stinging cells of jellyfish and their allies and the neural crest cells that are found in the vertebrates. In both cases, the cell-type involved is not found anywhere outside of the group concerned (respectively, the Cnidaria and the Vertebrata). Thus stinging cells must have arisen in the cnidarian stem lineage and neural crest cells in the stem lineage of the vertebrates.

Of course, nothing in evolution arises *de novo*: everything evolves from something else. So in one sense nothing is ever truly new. But in another, important, sense some things come much closer to being new than others. To try to illustrate this point, let's turn from the level of the cell to the level of large parts of the structure of animals. We'll look at two examples of novelties at this level, one in an arthropod, the other in a vertebrate: the venom claws of centipedes and the protective shell of turtles. But before getting into the details of these examples, it makes sense to discuss the non-novel or routine end of the spectrum of evolutionary changes.

To do this, let's return to the example of mammalian body size discussed in the previous chapter. This time, we'll focus on mice. Together with rats, these make up the subfamily called Murinae, with more than 500 species. Although all have small body sizes from our human perspective, the sizes of different species vary considerably. The familiar house mouse, which is one of the main model animals used for research into developmental processes, as we noted in Chapter 13, grows up to a maximum length, excluding the tail, of about 10 centimetres. The African pygmy mouse, as its name suggests, is smaller, while the brown rat is considerably larger – up to 25 centimetres in length.

As the various lineages of this subfamily of mice and rats radiated out from the stem species of the group, one of the kinds of evolutionary

FIGURE 23.1 The result of body-size evolution in the subfamily Murinae (mice and rats), as illustrated by a particular pairwise comparison. The house mouse and the brown (or Norway) rat are shown here at approximately the same scale. The latter is of course much larger than the former. The evolutionary changes in body size that have occurred in the lineages that led to each of these species have probably all been based on the standing variation that exists within most populations most of the time.

change that took place was body-size change. It's clear that there have been increases in some lineages, decreases in others (Figure 23.1); and that the magnitudes of these increases and decreases have been quite variable. Of course, other evolutionary changes have also taken place, both in externally visible characteristics, such as fur colour, and in internal features too, such as the nature of the stomach (related to changing diets). But we'll concentrate here on body size.

Generally speaking, when evolution produces a larger mouse from a smaller one, all its features are scaled up. Typically, a larger mouse will have longer forelegs, and longer hindlegs too – these showing co-variation, as we saw in the previous chapter. This is not to say that changes in shape don't occur – they do – but they are relatively minor.

Such evolutionary changes in the size of mice provide a good example of what I call routine evolution. We know that individual species, and indeed particular populations of individual species, exhibit variation in body size, some of which is due to genetic differences among individual mice. So it's not hard to see how evolution can produce larger forms from smaller ones, or vice versa. And of course this kind of evolutionary process is not restricted to mice; it's quite general, and can be applied to the size of any animal or the size of one of its constituent parts.

Since our two examples of evolutionary novelties are going to involve centipedes and turtles, it's worth mentioning that these two types of animal also exhibit routine evolution in body size and other characters, just as mice do. So we're not dealing with a scenario where some animals evolve in a routine way while others evolve in some magical way that produces novelties. Far from it. Not only are the novelty-inducing changes not magical, but the animals undergoing them will also be undergoing the more routine changes that are happening all the time in all animals everywhere. The situation is thus not either/or, it's one-or-both.

Just as there are many species of mice, so there are many species of centipede. At the most recent count, there were between 3000 and 4000 species in the world, and closer to the latter figure than the former. In terms of body-size variation, the smallest are less than 1 centimetre long as adults, the largest about 30 centimetres (a foot, if you prefer) – these are the largest of the tropical scolopendromorph centipedes, whose bite is fatal to small mammals such as mice and extremely painful for humans. It isn't known exactly what size the ur-centipede was, but it's a fairly safe bet that it was intermediate between these two extremes of today's centipede fauna. As centipedes have diversified, they, like mice and indeed most or perhaps even all groups of animals, have exhibited both increases and decreases in size. And importantly, these changes have been able to take place, at any point in evolutionary time, on the basis of the standing variation.

Let's now focus on centipede legs. These, in the same way as centipede trunks, vary in size among species: again, this is the result of routine evolution. However, something happened to a particular pair of legs in the centipede stem lineage some 400 million years ago (MYA), which was far from routine. I'm referring to the conversion of what was originally the first pair of walking legs into a pair of venom-injecting

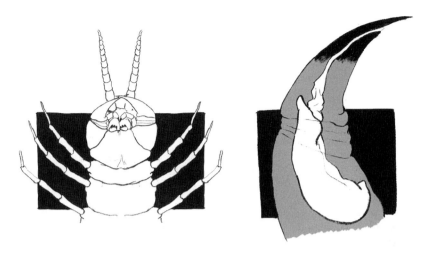

FIGURE 23.2 Left: the underside of a centipede head region, showing the pair of venom claws and the (very different) anterior pairs of legs. Right: an internal view – the venom gland and duct within a centipede's claw.

claws – the technical term for which is forcipules. A pair of these forcipules is pictured in Figure 23.2, both from an exterior perspective and internally to reveal the venom sac and duct. Three pairs of walking legs are also shown for comparison.

The external view shows that the forcipules have changed very markedly in shape from ordinary walking legs. Perhaps this aspect of the evolution of forcipules could be considered routine, because there is always variation in animal populations for shape as well as size. Although the amount of such variation in any species at any point in time is limited, given a sufficient period, consisting of millions of generations, evolution may be able to turn a thin walking leg into a stout forcipule that is wide at its base and tapers to a sharp point at its distal end.

What there is inside each forcipule, however, is far from a routine modification of what's inside a leg. Walking legs contain a lot of muscle tissue and not a lot else except things that are needed to work the muscles, which include anchorage points to regions of exoskeleton and nerves to stimulate the muscles into action. These things are also found inside forcipules, because they, like legs, are movable. But inside each forcipule is a superb venom production, transfer and injection system that has no parallel in legs at all. Where did this come from?

The short answer is that we really don't know. And this is the essence of an evolutionary novelty: something that seems to arise from nothing. There is a real challenge in trying to understand how such an evolutionary event is possible. For those who like their science neat and tidy, with all questions answered, this challenge is something of an annoyance. But to the *real* scientists, unsolved problems are the most exciting things, so the challenge is to be relished.

Not knowing the answer does not preclude the devising of hypotheses. Indeed, situations such as this are in a sense the realm of hypotheses. A hypothesis is essentially a suggested answer to an open question. One hypothesis for the origin of the venom system within a centipede forcipule involves *heterotopy*. This is the name given to structures undergoing an evolutionary shift in their position. Centipedes, along with most other animals, have many glandular systems in various parts of their bodies. Perhaps a mutation in a gene whose product acts to stimulate the development of a glandular system caused it to become switched on in a place where it was normally switched off – the first pair of legs. Perhaps its secretion was initially lubricatory, but further mutations resulted in the production of venom. Perhaps those centipedes with the most sharply pointed front legs and the most venom in the secretions from the glands within those legs were the most effective predators and were thus favoured by natural selection.

All of the above is indeed a hypothesis, one that has been put forward by the Irish-based French zoologist Michel Dugon in 2012 – hence the multiple use of 'perhaps'. It's a plausible enough hypothesis in many ways – such as how natural selection might favour centipedes with better venom-injection systems. But there's one aspect of it that would be rather difficult to confirm. The hypothesis, as framed above, would require an initial mutation to occur that would result in the production of glandular systems where they had not previously been. Such mutations are not likely to be part of the typical standing variation. So the evolutionary change that they help to initiate may be dependent, in a sense, on the right mutation coming along. Maybe so, but that's a hard claim to test.

Let's now turn to the other example of an evolutionary novelty that we're going to look at in some detail: the protective shell of turtles and tortoises. These animals form a monophyletic group (Chelonia). Its stem species probably lived just a few million years before the oldest known

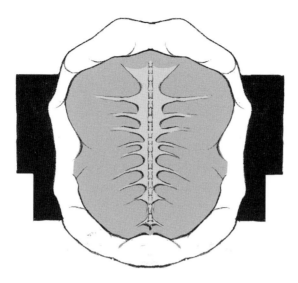

FIGURE 23.3 An internal view of the dorsal part (the carapace), of a turtle's shell, showing that the vertebrae and ribs are fused into the shell – with the result that the bones of the shoulder girdle, such as the scapula (shoulder blade), must be within the ribcage rather than outside it as they are in all other vertebrates.

fossil turtle, which dates from around 220 MYA. At present there are about 250 species in the group – this includes both the aquatic forms (turtles) and the terrestrial ones (tortoises). The body lengths of adult turtles and tortoises range widely – about 20-fold, from about 10 centimetres to about 2 metres. Although the group's monophyly is not in doubt, exactly where it fits into the tetrapod evolutionary tree is not clear. But what *is* clear is that the Chelonia must have arisen from a shell-less ancestor.

This means that the shell is an evolutionary novelty. It's not just the shell that's novel, though – the layout of the trunk skeleton is also novel in that the shoulder girdle, which is normally outside the ribcage, is inside it in turtles. These two things are connected. A turtle shell consists of two parts – the flattish ventral one (plastron) and the more domed dorsal one (carapace). A turtle's vertebrae and ribs are fused into the underside of the carapace (Figure 23.3), and thus the shoulder girdle has to be inside these structures. We are so used to the shoulder girdle being outside the ribcage that we normally don't give this layout a second thought. Each of us can feel our own scapula, or shoulder blade, and when we look at skeletons of any vertebrates other than turtles, whether dogs or tyrannosaurs, we see the same arrangement. Only in the Chelonia has the shoulder girdle been internalized.

The problem of understanding how such an evolutionary change can occur is similar in some ways to the problem we faced when examining the origin of the novel venom claws of centipedes. In both cases it would seem that what has happened cannot have been based on the typical standing variation found within species. There are no species of vertebrate in which some individuals have their shoulder girdle outside their ribcage while others have it inside. Also, in both cases, heterotopy – an evolutionary shifting of position – is involved. But the turtle novelty seems even more difficult to explain than the centipede one.

However, this is a case in which taking an explicitly developmental approach may help, as demonstrated by the American developmental biologist Scott Gilbert and his colleagues in 2001. If we think only of adults, then the origin of the shell and the internalizing of the shoulder girdle seem to be insurmountable challenges. But if we think of turtle embryos and the way in which they grow and develop, a possible solution emerges.

Turtle embryos have a structure called the carapacial ridge – essentially a region around the periphery of what will become the dorsal part of the shell, i.e. the carapace. This ridge plays a role in deflecting rib growth laterally. Thus in a turtle embryo, in contrast to what might be called a normal vertebrate embryo, instead of the ribs (which start dorsally) curving downward and then inward, they grow outward, attracted towards the carapacial ridge by developmental signals. Also, at certain stages of embryonic development in turtles, elements of the shoulder girdle, such as the scapula, lie anterior to the ribcage – so that if the ribs are deflected out laterally there is no collision between these and the shoulder girdle. And later, the shoulder girdle can be enveloped by the ribcage as the former moves a bit more posterior relative to the latter. So a change that looked impossible taking a three-dimensional view (adult structure) becomes possible when taking a four-dimensional (developmental) view.

There is, however, a big difference between possible and probable. It's worth exploring this point in relation to the comparative rarity of novelties. We've concentrated on just two particular examples of novelties here in order to keep things manageable. There are, however, many other examples. These include both well-known and little-known ones.

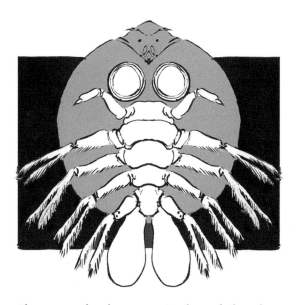

FIGURE 23.4 The two round suckers, or suction discs, of a branchiuran crustacean. These animals are parasites which use their suckers to attach to their hosts. Studies of the development of these structures show that they have been modified, in evolution, from leg-like mouth appendages (the first maxillae) in ancestral forms.

Everyone knows about the evolutionary novelty that we call *feathers* – found only in birds and some dinosaurs. But few people, other than specialists in the field, know of the *suckers* that are used by a group of parasitic crustaceans (branchiurans) to attach to their hosts. Developmental studies published by the Japanese biologist Tomonari Kaji and colleagues in 2011 have shown that these unusual structures (Figure 23.4) form from a pair of mouthparts that in other crustaceans resemble little legs.

   If it were possible to agree on exactly how to define an evolutionary novelty, we could in theory enumerate *all* cases of novelty origins rather than just give selected examples. Although we can't yet do this, it's clear that while novelties are numerous they are rare in comparison to what I have called routine evolutionary changes, such as changes in size and shape, which are ubiquitous in the animal kingdom, and indeed in the living world as a whole. At any one point in evolutionary time, all lineages are undergoing routine changes, while just a very few are experiencing the origin of a novelty.

Another way to put this is that the origin of an evolutionary novelty is possible but very improbable. And it seems obvious to ask why this is the case. Why is evolution normally conservative and only rarely radical in terms of what it does to animal design? The answer to this question is almost certainly to be found in the realm of the variation that Darwinian natural selection acts upon. Any evolutionary change that can take place based on the kind of variation that exists in most or all species most or all of the time – the standing variation – is highly probable. Or, to put this in another way, we should expect it to happen often. But evolutionary changes that cannot occur until some unusual variant arises through mutation would be expected to be comparatively rare – exactly what is found.

I said at the start of this chapter that evolutionary novelties are notoriously hard to define. That's why we have proceeded to look at examples rather than to discuss the various attempts at definitions. And indeed it may be that no definition is possible because instead of there being two categories of evolution – the routine sort and the sort that gives rise to novelties – there is a continuum of *degree* of novelty. In other words, if we consider the ways that evolution can act to repattern the development of an animal, perhaps size change is at one end of this continuum and things like the origin of the chelonian shell or of the centipede venom claw system are at the other. The further we go towards the novel end of the continuum, the harder it is to explain the evolutionary changes that are involved, but the more satisfying it is if we manage to do so.

24 Human origins and evolution

Earlier in the book (Chapter 7), we noted that the evolutionary relationships among the great apes are now well established. The closest living relatives to humans are the chimps. The human and chimp lineages split from each other a little over 6 million years ago (MYA). Although both chimps and humans have changed from our last common ancestor (LCA), we have changed more than they have. The most striking changes in humans are the adoption of an upright, bipedal posture, the marked reduction in body hair, and our increased brain size, with its corollaries of enhanced intelligence and behavioural repertoire.

Here, I'm going to focus on evolution of the brain and intelligence. It's clear that in this area humans have made something of a quantum leap from the human/chimp LCA, while chimps have not. Let's adopt an unusual approach to this topic. The object shown in Figure 24.1 is the *Huygens* lander – a small spacecraft that was launched from a larger one (the *Cassini* satellite in orbit around the planet Saturn). The *Huygens* craft has the distinction of having landed further away from Earth than any other object designed and made by humans – to be specific, it landed on the surface of Saturn's largest moon, Titan.

Huygens landed on Titan on 14 January 2005. It sent information in the form of radio signals to *Cassini*; the information was then relayed back to Earth. This information included a photograph of Titan's surface – which revealed a sandy-looking landscape strewn with stones, as shown in Figure 24.1.

It's worth considering all the things that went into making the successful landing of *Huygens* on Titan possible. These include: mining metal ores, purification of metals from these, highly precise engineering, astronomical knowledge relating to the dynamics of the solar system, manufacture of a massive launch rocket (of the Titan IV series) with sufficient power to overcome the Earth's gravity, understanding of the electromagnetic spectrum including the component of it that we call radio waves, design of transmitters and receivers, and a huge amount of computer software. Many of these things are underlain by considerable knowledge of the laws of mathematics and physics.

FIGURE 24.1 The *Huygens* lander, a spacecraft measuring only about a metre in diameter. This craft has the distinction of having landed further away from Earth than any other human-made object – on Saturn's moon Titan. Technology such as this is only possible because of our large brains. However, there was a gap of about 200,000 years between attaining our current brain size and achieving feats of planetary exploration.

We know that chimps are highly intelligent animals. However, they are a long way from launching space-probes. So the difference between us and them, as it stands in the early twenty-first century, is huge. And this difference is ultimately due to our different brain sizes. However, the connection is not as direct as it might at first seem. If we think back over human history using appropriate sizes of mental leap, we see an interesting picture. One hundred years before *Huygens* touched down, there were no spacecraft; indeed the first flight by a powered aircraft had only just taken place. One thousand years earlier, engineering was restricted to the manufacture of simple objects such as swords and coats of armour. And 10,000 years earlier (about 8000 BCE), there was no metal-working: the Bronze and Iron Ages lay in the future.

Now contrast these dates with another: 200,000 years ago. It seems likely, from fossil evidence, that by about then we humans had attained our current brain size of approximately 1350 cm^3. But with no understanding of science or mathematics, and little in the way of manufacture beyond the crude shaping of stones, the differences between the activities of humans and chimps would not have been particularly striking – for example to an extraterrestrial who visited Earth at that time. So we evolved a large brain, and we became able to launch space-probes, but the gap in time between these two events was about 200,000 years. One of the things this gap tells us is that our large brains have enabled a process of cultural transmission to build later knowledge upon the base of earlier knowledge, with the result that our entire knowledge-set has increased, and indeed may have done so at an increasing rate: certainly the technological achievements that have flowed from this knowledge seem to have increased exponentially.

The exact brain size of the human/chimp LCA is unknown, but it was probably not too different from the brain size of present-day chimps – about 400 cm^3. If this is correct, brain size in the chimp lineage has flat-lined over the last 6 million years. But in the human lineage brain size has more than tripled.

The rate of increase in our brain size over time seems not to have been linear. Rather, it seems to have been slow at first and then to have sped up. This pattern has been deduced from studying fossils of the hominin species that were probably ancestral to humans. (Hominini is a group that includes humans and our closest (extinct) relatives; it is narrower than, and nested within, the family Hominidae, which includes also the other great apes.) However, exactly which of the many hominin species that are now known *were* our ancestors is uncertain. The picture has kept changing with new fossil finds. Perhaps it has now stabilized, but then again perhaps such a conclusion is premature.

What we can say with near certainty is that about halfway between the separation of human and chimp lineages and the present, thus about 3 MYA, the proto-humans that would eventually lead to *Homo sapiens* were of a species of *Australopithecus* (meaning southern ape). At this point in time, all hominin species were in sub-Saharan Africa – hence the 'southern'. The species most likely to have given rise to the later *Homo* was *Australopithecus afarensis* (Figure 24.2). The brain size of adults of this species has been estimated as being about 450 cm^3. A million years later, hence about 2 MYA, our ancestor was probably the species called *Homo habilis* (handy man), with an estimated adult brain size of about 650 cm^3 (Figure 24.2). Coming forward another million years, our ancestral line may have been *Homo erectus* (upright man) with an adult brain size of about 900 cm^3. Given these figures, the increase in brain size per million years over the last 3 million years has been: + 200, + 250, + 450 cm^3.

There are at least two caveats to bear in mind. First, there are now about 10 recognized species of *Homo* and about five of *Australopithecus*. The route through these taken by our lineage may have been different to

FIGURE 24.2 The heads of three hominin species: *Australopithecus afarensis*, *Homo habilis* and *Homo erectus*. The approximate brain sizes of these were, respectively, 450, 650 and 900 cm^3. They represent an evolutionary time series; they also *may* represent a series of ancestors and descendants.

that suggested above. For example, *H. erectus* may have been a side-branch rather than part of our line of descent. Second, even if the ancestral route suggested is correct, brain sizes in all fossil species are estimated from imperfect evidence – usually fragments of skulls from which the full skull is recreated and the brain size then estimated from the size of the cranial cavity. So the figures given should be interpreted with care. Nevertheless, the accelerating-increase pattern seems to be fairly robust.

There is another factor to be considered here: evolution of body size. If intelligence was determined simply by the size of the brain, then the brightest animals of all would be sperm whales – with a brain size of about 8000 cm^3 – about six times that of humans. But what is in fact more closely related to intelligence is the ratio of brain size to body size. Humans have a very high such ratio, though strangely not the highest of all. Mice have a roughly similar ratio, and diverse other animals, including some birds and some insects, have even higher ones. Our uniquely high intelligence requires yet another factor to be taken into account.

This factor is the composition of the brain and the issue of which of its parts have increased in size the most in any particular lineage. The part of the brain that is associated with thinking is the cerebral cortex. And, as you might expect, this is the part that has undergone the biggest increase in the human lineage.

Finally, you may have noticed that, in giving various brain size figures above, I often qualified the figure by stating that it was the adult one. It's worth recalling that in this respect, as in many others, development (both embryonic and subsequent) recapitulates evolution. An early human embryo – for example at the gastrula stage – has no brain at all. Brain development begins at the next stage – neurulation. It then speeds up during organogenesis – hardly surprising, as the brain is one of the many organs after whose generation this phase of embryogenesis is named. Late embryonic growth produces a considerable further increase in brain size. Indeed, newborn humans have bigger brains for their size than do adults – the head grows more slowly than does the rest of the body during post-embryonic development.

Exactly when, in the history of the evolutionary lineage that led to modern humans, the creatures concerned became sufficiently similar to us to warrant being classified as belonging to our species – *Homo sapiens* – is not clear. Current thinking places this point in time at approximately 200,000 years ago. This is the figure I used above when estimating the time from the evolution of large brain size to the launching of space-probes.

This period of time saw several notable events relating to human geographical distribution and to human society and culture. It's thought that *Homo sapiens* spread out of Africa through the Middle East to the rest of the world sometime between 100,000 and 50,000 years ago. This was not the first exodus of a *Homo* species from the African continent – for example, the famous Chinese fossils known as Peking Man belong to the species *Homo erectus*. But such spreads were followed by extinctions.

There seems to be little doubt that human populations of 50,000 years ago, both those that remained in Africa and those that ended up in other continents, were of the hunter–gatherer type. That is, their food consisted of wild animals and plants. To get from that sort of society (if indeed the word is appropriate) to the sort found today, several transitions were needed. These can be labelled, perhaps rather crudely, as revolutions. The first of these was the agricultural revolution, which began about 10,000 years ago in the Middle East. The second was the industrial revolution, which took place between about 1750 and 1850 in parts of Europe and, later, the United States. The third was the digital revolution, beginning around 1940. The British code-breaking machines of Bletchley Park, developed in the Second World War, were among the first computers (though this term can be defined in many ways). The establishment of the Internet and the World Wide Web followed a

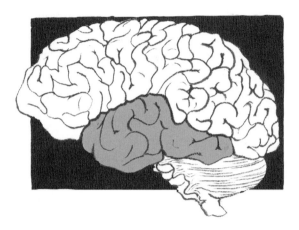

FIGURE 24.3 The brain of a present-day human, with the temporal lobe of the cerebral cortex shaded. This lobe is known to be important in relation to language.

few decades later (1980s and 1990s). The digital revolution is still with us and shows no signs of slowing down.

The growth of human thought does not connect very easily with the above revolutions. Notable progress in mathematics was made in ancient Greece. Progress in understanding the solar system came with Copernicus, Galileo and others in the sixteenth and seventeenth centuries – pre-dating the industrial revolution. Darwinian evolutionary theory dates, of course, from the nineteenth century.

The evolution of a large brain was almost certainly a prerequisite for growth in human technology – but, as already noted, there was a long gap between the two. Hence the benefits of the knowledge and technology of, say, the last 500 years cannot have been the reasons for natural selection favouring large brain size a long time before then. Evolution has no way of planning ahead – in this respect the dubbing of Darwinian selection as *The Blind Watchmaker* by English evolutionary biologist Richard Dawkins in his 1986 book of that title was very apt.

Why, then, did we evolve such a large brain? Strangely, this question is hard to answer. Hypotheses include selection related to one or more of several factors: tool-making and hunting; social interactions and language. Interestingly, one of the parts of the human brain which has increased in size the most over evolutionary time is the temporal lobe of the cerebral cortex, which is important in language processing (Figure 24.3). Of course, the exact nature of the selection that caused the evolution of our large brain may have changed over the course of time; if so, this is another example of exaptation (see Chapter 21).

The question arises, in connecting this chapter with the previous one, of whether the evolution of a brain size of 1350 cm^3 from one of about 400 cm^3 should be viewed as routine evolution or the origin of an evolutionary novelty. Increases (and decreases) in size would generally be put into the former category. So, too, would changes in shape – such as the expansion of some lobes of the cerebral cortex more than others. Therefore, looked at anatomically, there seems to be a persuasive case for regarding human brain evolution as routine. However, looked at in terms of the resultant possible behaviours, routine hardly seems to be the right word. So the situation appears to be one of novel behaviours arising from routine evolution of anatomy. This conclusion adds another dimension to the argument developed in the previous chapter that there is not a neatly defined category that we can call an evolutionary novelty, but rather a spectrum of the *degree* of novelty.

It's worth noting that the two original architects of the theory of natural selection, Darwin and Wallace, differed in their views of the reasons for the evolutionary increase in brain size in the human lineage. While Darwin saw natural selection as being responsible, Wallace thought that natural selection was insufficient to explain the origin of human consciousness and our higher mental faculties, including the ability to think in the abstract way that is required for a variety of pursuits, most notably, perhaps, mathematics. Wallace believed that some spiritual factor, not explicable in terms of natural selection, or indeed any natural agency, must have been involved in the production of the human mind. We'll look at this difference between Darwin's and Wallace's views further in the final chapter.

In current evolutionary biology, the consensus is that Darwin was right in relation to this matter, Wallace wrong. There is a reluctance among biologists to accept any appeal to non-natural causes, because that would be against the ethos of science. I would wholeheartedly support this reluctance. However, at the same time, I would feel more comfortable with the Darwinian approach to the evolution of human consciousness and abstract thinking ability if we had more satisfactory hypotheses regarding their selective value. Perhaps such hypotheses will be forthcoming in the not-too-distant future.

FIGURE 24.4 Normal (left) and sickle-shaped (right) human red blood cells. The shape of these cells is determined by the nature of the haemoglobin protein within them. The normal cells are roughly the shape of car wheels. The sickle cells are distorted in shape because their abnormal haemoglobin forms quasi-crystalline rods.

It's important to realize that gross anatomical measures such as the total size of the brain or the relative sizes of its constituent parts effectively gloss over a lot of detail at lower levels of organization – for example that of the cell. The number of nerve cells in the adult human brain has been estimated to be approximately 100 billion. With this many cells, the number of possible interconnections between them is almost infinite. But this latter – unknown – number would be a better guide to brain function than is any simple anatomical measure.

Moving from brain cells to blood cells, there is a good example of natural selection acting in certain human populations on one of the genes that produce our oxygen-transport protein haemoglobin, which is the main protein found in red blood cells. In adult humans the protein is made by two genes – the haemoglobin alpha and beta genes. A mutation in the beta gene causes a tiny change in the resultant protein sequence (just one amino acid out of 146 in the beta-chain is altered), but that tiny change has a significant effect: it causes the haemoglobin to form quasi-crystalline rods that distort the shape of the cells from their normal disc shape to a sickle shape (Figure 24.4). This reduces the cells' oxygen-carrying capacity, resulting in anaemia and often in premature death. Thus the disease is called sickle-cell anaemia.

This disease is rare in most parts of the world but common in areas of Africa where there is malaria; and there's a connection between the two. To see this connection we need to think not in terms of genes but of genotypes. Where there are two versions of a gene there are three genotypes – as in the case of the body-length gene discussed in Chapter 22. In relation to the sickle-cell disease, two copies of the mutant haemoglobin beta gene may be present in a given individual – in this case the individual is anaemic. If, in contrast, both copies of the gene are normal, then of course the individual concerned is not anaemic. Individuals with the third genotype (the heterozygote) do not suffer from anaemia because they have sufficient normal haemoglobin but they are carriers of the mutant gene.

Since there is selection against individuals with anaemia, why does the disease not simply die out? In other words, why does natural selection not act to eliminate the mutant gene from the populations that have it? The answer lies in the fact that heterozygotes have resistance to malaria. So in malarial regions it can be the case that heterozygotes are the fittest of the three genotypes. This means that the selection is not directional but rather balancing, and acts to retain the mutant gene in the populations concerned. We encountered balancing selection much earlier in the book – in Chapter 5, where we saw an example in snails. It's probably quite common, and this should give us cause to reflect on the complexity of natural selection and the fact that it can be an agent for stasis as well as an agent of change.

It's important to realize that selection of the directional sort is also taking place in current human populations: we have not stopped evolving. One possible form of selection that may be taking place in first-world countries is selection for faster reaction times. The reason why I suggest this is as follows. In many of these countries the biggest killer of teenagers and young adults is road traffic accidents. One of the many factors that contribute to whether an individual falls victim to such accidents is reaction time. On average, individuals with faster reaction times to indications of imminent danger will have a lower probability of dying. Such small differences in the probability of mortality are the stuff of natural selection.

It is also possible that humans are evolving increased height. Anyone who has visited a house built several centuries ago, such as Anne Hathaway's cottage (the childhood home of Shakespeare's wife) will notice how low the tops of the doorframes are. English people were shorter, on average, then (in the sixteenth century) than they are now. Increases in height have occurred in parallel elsewhere.

Not only is it not clear what the selective advantage of increased height might be, it's not even clear that there was one. What I'm getting at here is that there is a possibility that height increase over the last few centuries may not be inherited. Height is one of the many characters in which the variation among individuals has both genetic and environmental – including dietary – components. It's possible that dietary improvements, rather than Darwinian selection, have caused the height increase.

Looking to the future, how might humans be expected to evolve? Will we, for example, undergo further increases in brain size? Such questions cannot be answered, because evolution is largely an unpredictable process – partly because the ways in which the environment will change are themselves unpredictable. So speculation about evolution in the future is the stuff of science fiction, not science. Perhaps in the distant future humans will have enormous heads to accommodate enormous brains, like those many fictional aliens we see depicted – but perhaps not. This is one of those areas in which we simply have to admit that we don't know the answer.

25 Animal plasticity

We can generalize the point made in the last chapter about the role of environmental factors in determining human height to the role of these factors in determining the body sizes of most if not all animals. Indeed, we can generalize further, because environmental factors play a role in determining the values of many characters in most or all animals, including those that are unrelated to size. In other words, the genes do not completely determine the course of development; rather, this course is set by a mixture of genetic and environmental influences, and the interactions between them.

There are several ways of describing the effects of environmental factors on development. One way is to say that a character such as body size upon whose value there is an environmental influence has only a partial *heritability*. Another is to say that there is an element of *plasticity* in the character concerned. More complete terms for the latter are phenotypic plasticity or developmental plasticity, the second term being preferable for reasons already given in Chapter 22.

Plasticity extends beyond the minor, quantitative environmental effects usually seen in characters such as body size. There are situations in which the environment can have a major effect on the route that development takes. These situations often involve the existence of two or more qualitatively different types of animal in the population: something that is called *polyphenism*. (If there are two or more discretely different types present due to genetic reasons, we call this, instead, a *polymorphism*.)

It's interesting to consider the commonest kind of 'existence of two discretely different types of animal' in a population – two sexes – against this conceptual background. In humans, and mammals in general, the existence of males and females in a population is a special case of polymorphism rather than polyphenism, because the sexes are genetically determined. This is also true in birds. However, in some other vertebrates the existence of animals belonging to two sexes is a polyphenism because sex is determined by the environment. One of the groups in which this phenomenon has been much studied is the

FIGURE 25.1 The brood chamber (or nest) of a sea turtle, containing a large clutch of eggs. The embryos develop within a chamber for several weeks, the exact time depending on the species. The temperature in the chamber determines whether the embryos develop as males or females.

Chelonia – turtles and tortoises – which we looked at in Chapter 23 for a very different reason, namely consideration of their shell as an evolutionary novelty.

In many species of turtle, sex is determined by temperature; specifically, the temperature within the brood chamber. But what is this place? To answer that question it's necessary to digress to examine some aspects of the natural history of turtles. As we've noted earlier, the group Chelonia includes marine, freshwater and terrestrial forms – with the first two of these normally being called turtles, the third tortoises. In all three habitats, eggs are usually laid in a brood chamber or nest. The placement of this depends both on the broad habitat type and the exact species concerned. The adult females of many sea turtles make their way up a beach and excavate a brood chamber in the sand using their flippers (Figure 25.1).

The number of eggs and the time they take to develop also depends on the species. Egg number can range from fewer than 10 to more than 100. Time to hatching is typically between one and three months. During this time, the temperature in the brood chamber varies, both over a

24-hour cycle and from one day to the next. The sex of a turtle is determined most by the maximum points of the temperature cycle; and also more by the temperature in the middle part of embryogenesis than by the temperature at the start or end of the process.

In most turtles the female has a larger average body size than the male. Higher temperatures typically produce female turtles and lower ones males; whatever the exact mechanism underlying the effect of temperature on sex determination, the link with body size is probably important. This is thought to be the case because in crocodiles, where most or all species have temperature-dependent sex determination, higher temperatures produce males; and these are of larger body size than females. Thus the common factor seems to be that higher temperature produces the sex that has the larger body size. Whether this is always true though remains to be seen. Many species in both groups have not yet been studied in detail in this respect.

Of course, body size is not the only difference between male and female turtles. The reproductive systems are different in the usual way, organized to produce sperm in males and eggs in females; and some other features are different too – for example larger forelimbs in the male, which are used to clasp the female during mating. Interestingly, the structure of the penis of male turtles suggests that it is an independent evolutionary invention from the penis of mammals – if so, then this is another example of convergent evolution to add to the many we've already come across in other contexts.

There are many other examples of polyphenism in many different animals. Moving to invertebrates, there is a type of small crustacean inhabiting ponds and other freshwater habitats called a water flea (though it's not a flea at all; that's an insect). There are many species of water flea belonging to the genus *Daphnia*. Adults normally have a rounded head and a short tail spine, but when certain predators are present they develop a pointed helmet and a longer tail spine (Figure 25.2). It seems that the developmental system of the water fleas reacts to some chemicals in the water that derive from the predators.

This example shows that in some cases a polyphenism involves the presence of two or more forms in the same population but not necessarily at the same time. Another example of this is the seasonal polyphenism found in the wing eyespots of the species of butterfly known as the African brown. We looked at these eyespots in a different context in

FIGURE 25.2 Helmeted (left) and unhelmeted forms of the water flea *Daphnia*. The former is produced in response to predator-derived chemicals in the water. This is an example of a plastic developmental response that is adaptive. Although often described as a polyphenism with two discrete types, intermediates between the forms shown here are known to occur.

Chapter 22, where we saw that part of their basis was genetic. However, another part is environmental: populations of this species have more pronounced eyespots in the wet season than in the dry one.

One of the best-known types of polyphenism is that involving different social castes in the insect group Hymenoptera – which includes ants, bees and wasps. There are many variant arrangements regarding these castes and the hymenopteran society of which they are a part. The European (or western) honeybee *Apis mellifera* has been especially well studied in this and other respects, in large part because of its economic importance.

In honeybees, the difference between a (male) drone and a (female) worker or queen is genetic – males develop from unfertilized eggs, queens and workers from fertilized ones. However, the difference between a worker and a queen (Figure 25.3) is environmental. In this case, the environmental effect occurs through diet: a queen will develop instead of a worker from a larva that is fed almost exclusively with royal jelly, a sugar-rich secretion from workers. However, in addition to the

FIGURE 25.3 Queen (left) and worker honeybees of the species *Apis mellifera*.
Note the difference in size, with the queen being significantly larger. There are
also subtle differences in shape, such as the queen having a somewhat more
tapering abdomen. Queens and workers are determined not genetically but
environmentally, and in particular through the feeding regime of larvae, with
queens developing as a result of being fed with large amounts of royal jelly.

dietary effect, queens normally derive from larvae developing in special
cells in the colony that are known simply as queen cells.

In general a colony only has one queen. However, if the existing
queen dies, or becomes sufficiently old or diseased that her pheromone
output is reduced significantly, then workers begin to feed royal jelly
in large quantities to larvae in queen cells. The usual result is that
the first queen to develop to adulthood kills the other developing queen
larvae.

Colonies of *Apis mellifera* have a self-regulating ability in the sense
that when a nest or hive becomes overcrowded a new queen is produced.
Usually, the old queen and a swarm of several thousand workers will
leave and found a new nest, which is built by the workers using wax
secretions. Once the new nest and colony are established, the colony
will operate much as the parental one did until, when it too becomes
overcrowded, the process of founding a new colony begins again.

FIGURE 25.4 Queen and soldier termites; workers are typically similar in size to, or a little smaller than, soldiers. Here, the difference in size (and shape) between the queen and other forms is much greater than in honeybees. Nevertheless, it still has an environmental, rather than a genetic, cause.

The kind of social living exhibited by the honeybee is referred to as eusociality – a true society, if you like. This high degree of sociality is found not only in many other hymenopterans but in some other groups of insects too, most notably the termites, where it has arisen independently – yet another example of convergent evolution. In many termites there are more castes than in honeybees – for example, there is often at least one soldier caste. Also, the difference between a termite queen and worker (or soldier) in size and other features (Figure 25.4) makes the difference between honeybee queens and workers look insignificant.

We started this chapter by making a connection with the previous one in relation to environmental influences on human height in particular and on animal body size in general. In this context, the environmental effects concerned are generally small, helping to produce the continuous variation that is normally observed in body size between different members of the same species. Since then, we've been examining examples of polyphenism – where the environmental effects are greater and the resultant differences between forms are typically discrete rather than continuous. In the case of environmental sex determination,

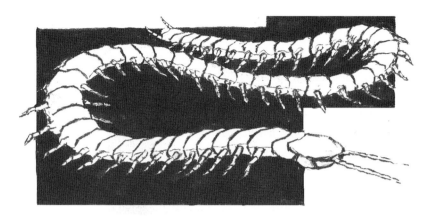

FIGURE 25.5 An adult female *Strigamia maritima* centipede. Individuals of this species have between 43 and 53 trunk segments. It is not yet clear whether the exact number matters. It may be that segment number is a selectively neutral character.

there are of course just two forms – male and female. In insect societies there are usually at least three castes, but not all determined by environmental effects – for example the drone honeybee is genetically different from other castes, as we noted. In general, polyphenism is restricted to situations in which the number embodied in the 'poly' is small – usually two, three or four.

To contrast with that, let's now examine a case where the number of forms can be rather more: the number of leg-bearing segments in centipedes of the coastal species *Strigamia maritima* (Figure 25.5). This number can be any of the following: 43, 45, 47, 49, 51 or 53. It's possible that individuals with 41 or 55 segments also exist but are very rare. (Even numbers of segments are not found; this is an example of developmental constraint – see Chapter 27 for more on this general subject area.) So the number of forms is at least six, and may be seven or eight. It's known from laboratory rearing experiments that the determination of segment number takes place by a combination of genetic and environmental effects. The main environmental effect is that of temperature: the higher the temperature experienced during embryogenesis, the greater the number of segments. This kind of effect is probably a general one in the group of centipedes that *Strigamia* falls within – the Geophilomorpha, or ground-loving forms, with more than 1500 constituent species.

This is a good point at which to connect developmental plasticity with evolution. There are two key questions here. First, is plasticity adaptive? And second, might plastic effects help us to understand the origin of evolutionary novelties? We'll take them in that order.

With regard to the adaptiveness or otherwise of plasticity, it's useful to contrast the *Daphnia* (water flea) and *Strigamia* (centipede) cases. In *Daphnia*, the developmental changes that are invoked by the presence of a predator are almost certainly adaptive. It's hard to believe that the chemicals secreted into the water by the predators produce the development of structures (helmets, spines) that just happen by coincidence to act as predator-deterrents. A more plausible hypothesis is that natural selection in the past favoured those *Daphnia* individuals whose developmental system reacted to predator-chemicals in this way. So there is a similarity here with the well-known form of plasticity found in many plants which involves growth being oriented so that the plant bends towards the main source of light.

A question that remains open in the *Daphnia* example is why natural selection should favour a plastic response rather than a fixed predator-deterring growth form. After all, predators come and go; surely it would be better to be generally defended against them rather than have to wait to develop defences after they arrive? Perhaps the answer lies in energy economy: why waste energy making pronounced helmets and spines if these are not needed? Anyhow, regardless of the answer to this question, the adaptiveness of the plasticity is not in doubt.

In contrast, there most certainly is doubt about whether it is better for a centipede like *Strigamia* to have more segments when the temperature is higher. It remains a distinct possibility that temperature simply affects the development of *Strigamia* in this way; and that natural selection does not act for or against it doing so because, if you have lots of similar body-segments, then whether you have a few more or a few less is unimportant to your survival. If this is so, then by definition the segment-number-enhancing effect is what we call selectively neutral.

Notice that the kind of selection under discussion here is more complex than that which is involved in cases of characters that are completely genetically determined. In the latter case, selection favours those individuals that have genes producing higher or lower values of the character concerned. But when a character is plastic, what selection must be doing, for example in *Daphnia*, is favouring individuals whose developmental systems produce a different outcome when there is a certain environmental stimulus.

If you peruse the literature on developmental plasticity, which is extensive, you will find some authors attempting to restrict the use of the term *plasticity* to those cases of environmentally induced developmental trajectories that *are* adaptive. This seems an unwise approach to me, because of cases like *Strigamia* in which the adaptiveness or otherwise of the environmental effect on development is unclear. And this is not a unique case in which there is a lack of clarity in this respect. There are many more. To illustrate this point, let's return to the issue with which this chapter started – environmental effects on animal body size.

One widespread form of plasticity is a negative relationship between average body size and population density. This has been studied in a very broad range of animals, from insects to fish. It has also been studied in plants. Here, we'll restrict our attention to the fruit-fly *Drosophila melanogaster*, in which this negative relationship has been studied in considerable detail.

As with most flies, the ecology of *Drosophila* adults is very different from that of their larvae. Populations kept in the laboratory are reared on a food supply that is designed to maximize larval growth and survival. Different larval densities can be engineered at the start of a laboratory culture by manipulating the number of larvae used and/or the amount of food supplied, since density is measured as larvae per unit food. Experiments comparing cultures started at different larval densities have shown that the higher the larval density the lower the resultant adult body size.

One interpretation of this finding is that the form of plasticity involved is inevitable rather than adaptive. In other words, if less food is available larvae will grow more slowly, pupate at a smaller size, and end up producing adults that are smaller. So the small adults produced from high-density cultures are not selectively favoured. Indeed they may be at a selective disadvantage, because it's known that smaller adult females lay fewer eggs than do larger ones. So perhaps we are simply looking, here, at a starvation effect.

However, there is an alternative possible explanation. Larvae need to reach a certain minimum size in order to pupate successfully. Suppose a species of fly evolves in which there are initially two genetic variants, one in which the larvae must always reach a fixed size to pupate and another in which the minimum size for pupation is variable, with this variation incorporating the ability to pupate at a smaller size when food is scarce. In conditions of restricted food, the latter variant may survive at the expense of the former. Better to produce small adults when times are hard than to produce none at all.

We will now turn to that other question about plasticity and evolution raised a few pages back: can plasticity help to explain the origin of evolutionary novelties? To examine this issue we'll stick with *Drosophila* fruit-flies. We'll look in particular at some experiments that were carried out several decades ago by the British biologist C. H. Waddington, who was based at Edinburgh University; this work was published in 1956.

You may recall from Chapter 19 on animal developmental genes that there are certain key genes that are involved in patterning the head-to-tail body axis, mutations of which can produce homeotic transformations (right-structure-in-wrong-place). One of these that we examined is the bithorax form in which the little flight-balancing organs normally protruding from each side of the third thoracic segment are transformed into wings. Strangely, what Waddington found was that in a population of normal flies (no bithorax mutation) the bithorax form could be produced by exposing eggs to a high concentration of ether vapour.

Interestingly, Waddington's experiments showed that only some flies emerging from cultures treated in this way were homeotically transformed. However, by selectively breeding from these Waddington was able to increase the frequency of the bithorax response to ether vapour. So there must have been some hidden genetic variation in the original population, which Waddington exposed by his ether treatment. Having exposed it, he was then able to use it as a basis for selection.

There's an important caveat here about the relationship between artificial and natural selection. Although the artificial selection conducted in the laboratory by Waddington and others can be used as a model of Darwinian natural selection, the forms made artificially fit by the experimenter may be very unfit in nature. This is the case with bithorax fruit-flies. So the importance of these experiments does not lie in telling us something about the bithorax form *per se*.

Rather, its importance lies in what it is telling us about ways of producing new variants in general.

Recall (from Chapter 22 on variation) that most animal populations have a certain amount of standing variation that is present throughout their history. This includes genetic variation for characters such as body size. But also recall that most animal populations have a limited amount of standing variation, so that when it comes to the evolution of a novelty such as the turtle's shell we have to invoke the occurrence of new and unusual variants. These might arise by mutation. However, such mutations will be rare, and when one occurs it will typically only affect a single individual. Mutations of this kind are usually lost from the population again (by the random process of genetic drift) before they can be spread by selection.

What Waddington's work shows us is that there is another way for new and unusual variants to arise: through plastic effects. And in this situation many variant individuals may arise at once. Although they have been rendered variant by an environmental stimulus, the particular individuals involved may have been predisposed to respond developmentally to that stimulus more readily than others in the population for genetic reasons. Thus we have a basis for selection to occur from what can be described as a better starting point than a single, random mutation. It's not yet clear whether such a process has been at work in the production of evolutionary novelties, but this is an interesting hypothesis that is well worth further consideration.

26 The nature of adaptation

Evolutionary concepts in general can be thought of as being primarily related to pattern in some cases and process in others. Darwinian natural selection can be considered to be the key concept related to process. With regard to pattern, the key concept can probably be said to be *homology*. Ironically, this term was invented by English anatomist Richard Owen in the 1840s, before publication of *On the Origin of Species* and without an explicit evolutionary interpretation. Owen recognized as homologies similarities that were in some sense true ones, despite obvious differences in form, such as the arm of a human and the wing of a bird (Figure 26.1). Others that were not thought to be true ones were not considered homologous but merely analogous – such as the tail of a salmon and that of a dolphin.

It soon became apparent, though, that what Owen was recognizing as homologies were structures that had a common ancestral form – the foreleg of an early land vertebrate in the case of the human arm and the avian wing. In contrast, similar but non-homologous structures arose convergently. A dolphin's tail arose as a novel structure when one lineage of mammals became aquatic. It is homologous to the tails of related mammals, notably whales, but not to the tails of fish.

Although I did not use the term homology when discussing evolutionary trees earlier (in Chapter 7), the use of monophyletic groups or clades, as formalized by Willi Hennig, is essentially an approach that defines taxonomic groups in terms of homologies. Darwin's notion of 'groups within groups' can then be seen as a pattern of nested homologies, as depicted in Figure 26.2 in the abstract form of sets and subsets.

Of course, pattern and process should not be viewed in isolation: they are related. The adaptation to life on land that produced legs from fins was responsible, along with the subsequent evolutionary radiation of lineages, for the creation of the set of tetrapods shown in Figure 26.2. And natural selection is the mechanism that produces adaptation. So we have a joined-up picture of evolution: selection produces adaptation; a newly adapted lineage will ramify due to further selection and speciation events; the result is an evolutionary tree

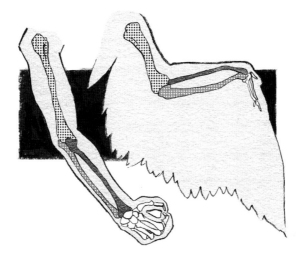

FIGURE 26.1 Homologous structures: the human arm and a bird's wing, showing that they have similar underlying bones. These have been modified in shape but not in basic layout. So, for example, the pattern of one long-bone (humerus) followed by two in parallel (radius and ulna) is common to both.

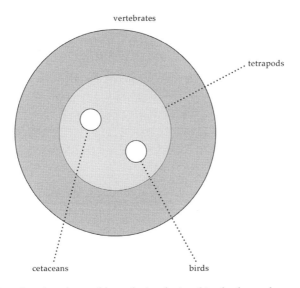

FIGURE 26.2 A series of nested homologies depicted in the form of a set diagram: birds and cetaceans (whales and dolphins) are shown nested within the tetrapods, which in turn are nested within the vertebrates.

growing through time; and at any given moment, such as the present, the structure of the overall group concerned can be viewed as a pattern of nested homologies.

The summary of evolutionary theory presented in the previous paragraph is fine as far as it goes, but it does not go far enough. All three key concepts in it – selection, adaptation and homology – are more complex than might appear from this or any other brief description. In the present chapter, I want to look in particular at some complexities involved in the idea of adaptation.

Adaptation is normally used to refer to the way in which animals (or organisms more generally) are suited to their environment. This can be looked at in various ways. First, if we simply consider some existing animal form, without explicit reference to its evolutionary origin, we might admire what appears to be its marvellous adaptation to its overall environment, or to some particular environmental factor. For example, a dolphin's sleek body, along with its fins and tail, seems wonderfully adapted to movement in water.

Second, we can sometimes observe adaptation happening, as in the famous case of the peppered moth, *Biston betularia*. Populations of this species living in urban areas of England (and elsewhere) in the 1800s experienced a new kind of environment in which surfaces on which moths could rest, such as tree trunks, were covered in soot. In such an environment, the normal pigmentation pattern of the wings (pale and speckled; hence 'peppered'), which had been well camouflaged and thus difficult for predators to spot against clean, lichen-covered tree trunks and branches, was no longer well camouflaged. The spread through these urban populations of a mutant, darkly pigmented, melanic form has for decades been a standard textbook example of an adaptation in the making (Figure 26.3).

Third, when several lineages branch out from a common ancestor, and the way in which they become different from each other seems to be related to the different environmental conditions prevailing in the different geographic places where the descendant species end up, we call this an adaptive radiation. The diversification of the various species of Darwin's finches on the Galapagos archipelago is one of the classic examples: see the 2008 book *How and Why Species Multiply: The Radiation of Darwin's Finches*, by the husband-and-wife team Peter and Rosemary Grant. In studies of these birds, particular attention has

FIGURE 26.3 The pale, speckled form of the moth *Biston betularia*, together with the melanic form. The pale form is better camouflaged against a background of lichen-covered tree trunk (top), whereas the melanic is better camouflaged against a soot-blackened trunk (bottom).

been paid to the form of the beak, as we saw in Chapter 22, and the way in which it adapts to the available food supply.

It might seem at this stage that adaptation is old hat. It's been at the core of evolutionary theory since Darwin put it there more than 150 years ago; it's been elucidated further by countless biologists since; and an account of it features in pretty much every textbook description of the mechanism of evolution. However, maybe the standard accounts are missing something, or at least downplaying it. Let's explore this interesting possibility.

There's one way in which the three examples of adaptation given above differ. We can call it, perhaps, environmental breadth. The streamlined bodies of dolphins are adapted to life in aquatic habitats – not just marine ones, but estuarine and freshwater ones too – there are several species of river dolphins. Given that aquatic habitats account for more than 70% of the Earth's surface, the adaptation of dolphins' bodies to life in water is in a sense a very *broad* adaptation. Along with that goes a lack of reversal of the body form concerned. Although some dolphins are a little less streamlined than others, all are very definitely streamlined when compared to their terrestrial mammalian ancestors and cousins.

The other two cases – the peppered moth and Darwin's finches – involve a lesser degree of environmental breadth and, associated with that, some degree of reversibility. In the moth, a reverse is not just possible, it has actually happened. Populations in the Manchester/ Liverpool urban area have been studied right up to the present day. In the last half-century or so, since the passing of the Clean Air Acts and the disappearance of widespread soot, the direction of change in urban populations has switched – the melanic moths have been declining in frequency from their high point of more than 90% to a current level of less than 50%. Perhaps they will disappear altogether after a few more decades.

We can generalize from this particular species of moth to all moths, and indeed butterflies too. Given that all species of Lepidoptera (the butterfly-and-moth group of insects) have a large expanse of wing surface, whose pigmentation pattern is exposed for all animals with eyes to see, natural selection has often changed the pattern, and for many different adaptive reasons. Some butterflies are conspicuously coloured because they are poisonous to bird predators; others have evolved to become mimics of the poisonous ones. Their lookalike pigmentation means that they get eaten less by birds because they are taken, incorrectly, to be poisonous too.

As we saw in Chapter 22, many butterflies have eyespots on their wings. When these are small, the selective advantage may be that birds peck at them rather than at the butterfly's head, with the result that the would-be prey escapes with nothing more than a nick in its wing. But where the eyespots are large the selection is different: it is thought that birds get startled by seeing a pair of very large eyes, mistakenly

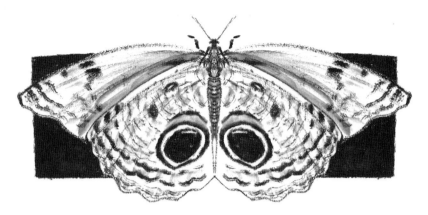

FIGURE 26.4 An owl butterfly. There are many species of these. They differ from the African brown butterfly which we looked at earlier in that the eyespots here are very much bigger. This alters the way in which natural selection acts.

taking them to be the eyes of a potential predator rather than merely the wing eyespots of a potential prey; this is the case with the owl butterflies (Figure 26.4).

The take-home message here is that the environments in which butterflies and moths are found differ from each other in terms of the visual background; they also differ in the array of predators present, and the degree of threat presented by these. Each environment presents very specific challenges, and hence natural selection acts in very different ways in different situations. The value of any one pigmentation pattern may be restricted to quite few environments; in other words, which form is best adapted changes from place to place and, as we saw with *Biston*, from time to time.

So far, this environmental-breadth dimension to the story of adaptation might seem interesting but hardly revolutionary. However, our perspective may perhaps change if we take this dimension right to its broadest end. What I'm getting at here is that some evolutionary changes might be such that they are beneficial in all, or almost all, environments. If this is so, then it becomes debatable whether the term adaptation is appropriate at all. Personally, I don't think it is; rather, a term such as 'generalized improvement' would seem more appropriate.

We've encountered some examples of generalized improvements before: a through-gut in an early bilaterian, and jaws in an early vertebrate. Let's now take the jaw example a bit further. Some examples of

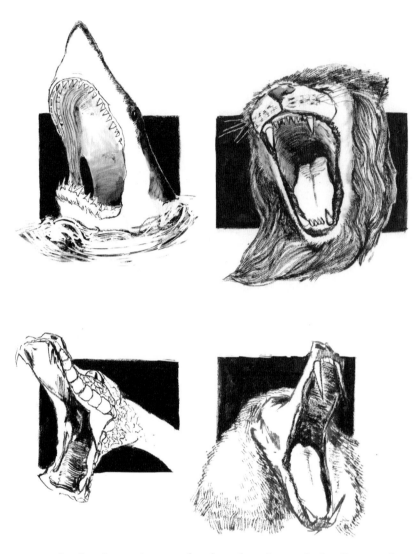

FIGURE 26.5 Four impressive examples of vertebrate jaws: a shark, a lion, a snake and a baboon. Having evolved, jaws are useful in so many different environments that, although they become evolutionarily modified in different lineages, as can be seen here, there is no lineage in which they have (at least yet) been lost.

jaws are shown in Figure 26.5. As can be seen, they are impressive, and in many cases fearsome, structures. We saw previously (in Chapter 20) that the very first vertebrates lacked jaws and had instead a roundish mouth – the condition seen in today's lampreys and hagfish. But the number of species belonging to these two groups is now tiny.

They comprise less then 1% of all vertebrates. The other 99% and more have jaws. The name for this larger group is rarely used in the more popular scientific literature, partly because it is an awkward word, and partly because it has an awkward status. In relation to the latter, recall that our own phylum is Chordata, while Vertebrata is a subphylum. Since familiar large groups within Vertebrata are classes – Mammalia and Aves for example – where does that leave the jawed vertebrates? Their group name is Gnathostomata (meaning jawed mouth). So this could be a sub-sub-phylum, an infra-phylum or a super-class – all of which are ungainly categories. But, while Gnathostomata may have a problematic taxonomic rank, it is a hugely important group of animals.

In Chapter 20, we saw that jaws arose from the skeletal supports for the first pair of gills. We also noted that the initial enlargement of these may well have been driven by selection to make the gills work better, and that only after the initial steps did the reason for the selectively driven changes alter to favouring an improved form of eating as opposed to an improved form of breathing. Furthermore, in Chapter 21 we noted that a structure which evolves first for one reason and then for another is said to have been exapted – yet another twist in the story of adaptation.

But here I don't wish to elaborate on the idea of exaptation, which is generally accepted now. Rather, I wish to consider the view that the evolution of jaws, and in particular the second stage of the process in which improved eating ability was the driving force, was not an adaptation at all. In my view, adaptation implies an improvement that is at least to some extent environment-specific. Even when the degree of specificity is low – as in the case of a fusiform body that's advantageous in all aquatic habitats – it's still specific to a degree, in that there are other environments (terrestrial ones) in which it would be a disadvantage.

In contrast, it can be argued that, for a vertebrate, jaws are an advantage everywhere. They thus represent a general design *improvement* rather than an environment-specific adaptation. And this argument can be extended from jaws to other features – especially those that are also involved in the stem lineages of high-ranking taxonomic groups such as phyla and super-groups of phyla. Indeed, you may recall my labelling, in Chapter 18, the evolution of a through-gut (Bilateria) and of a tough cuticle (Ecdysozoa) as being generalized improvements rather than environment-specific adaptations.

Let's return to vertebrates and consider the origin of the internal skeleton that makes them what they are: the cranium, the spine, and the (more variable) appendicular skeletal supports, such as fin-rays or leg-bones. Having these things – as opposed to having some particular version of them – is probably advantageous in all environments. Thus their evolutionary origin was, like that of jaws, a general improvement rather than a specific adaptation. Perhaps the origins of animal phyla and super-groups of phyla *usually* involve such generalized improvements.

The big issue here is the scale-dependence or scale-independence of evolution. In the year 2000, two linked papers were published in the journal *Evolution & Development*, one arguing for scale-dependence, the other for scale-independence. The latter, written by Armand Leroi, was simply titled 'The scale independence of evolution'. The other, written by Douglas Erwin, was putting forward the counter-argument and could have been titled 'The scale dependence of evolution'. In fact, it was titled 'Macroevolution is more than repeated rounds of microevolution'. Both the actual and the possible title of Erwin's article are fine, but I'm trying to avoid the use of microevolution and macroevolution here because their definitions are problematic, with some authors using a third term, megaevolution, for the biggest-scale changes.

Leroi noted in his introduction that his and Erwin's task was difficult because the issue of scale-dependence versus independence had "exercised the finest minds in evolutionary biology generation upon generation since *The Origin*". He went on to summarize the nature of the issue as follows. He described it as "the debate over whether or not the grand sweep of evolution visible to the comparative anatomist and palaeontologist can be explained by the minute evolutionary effects visible to the naturalist and experimentalist." To connect this with the examples used above, the origin of the vertebrate jaw and its spread at the expense of a jawless design is in the realm of the grand sweep of things that palaeontologists deal with, while the spread of melanic moths at the expense of the original pale and speckled form within the species *Biston betularia* is in the realm of things that can be studied by naturalists.

As is apparent from their titles, Leroi and Erwin took opposing positions in relation to this crucial debate. It's still possible, more than a decade later, for biologists to take either of these opposing positions.

However, let's see what emerges from taking the approach that I started with here – an approach that focuses on the degree of environmental breadth of an adaptation.

An important aspect of this approach, as noted earlier, is the degree of reversibility of the evolutionary change concerned. The broader the range of environments in which any particular change represents an improvement, the lower is the probability of a reversal.

However, while it's possible to find cases of broad-scale adaptations that have never been reversed, such as the streamlined bodies of cetaceans, there are plenty of cases in which broad-scale changes *have* been reversed. Indeed, as we have seen, the cetacean body form is itself the result of, among other things, the reversal of the fin-to-limb transition that produced the land vertebrates. Also, although no losses of the vertebrate endoskeleton have taken place, losses of bilateral symmetry have occurred, most notably in the echinoderms – this is essentially a reversal of the radial-to-bilateral symmetry change that produced the urbilaterian.

Consideration of the issue of evolutionary reversibility does not lead to a resolution of the debate in favour of scale-dependence. Rather, it refines the nature of the debate, and in particular one side of it. The argument for scale-independence remains much as it was. However, the argument for scale-dependence now goes as follows. Generalized design improvements tend to persist over long periods of evolutionary time; they are retained in all (or most) descendant lineages, in each of which the new feature (e.g. a pair of jaws) is modified in different ways. In contrast, more environment-specific adaptations tend to be reversed or lost in various descendant lineages. The overall history of a major group, such as the vertebrates, or, within them, the gnathostomes, includes the origin of generalized improvement. In contrast, the history of one lineage over, say, a few million years usually does not.

Let's relate this view back to the diagram of homologies shown in the form of sets (Figure 26.2). The origin of a cranium and vertebral column came first in evolution and represented a design improvement in all environments. These features have never been lost. The origins of walking legs (tetrapods) and of flying wings (e.g. birds) are more environment-specific (though still rather broad). Both have been lost – or reversed, if you prefer – as for example in cetaceans and penguins respectively.

Having progressed the scale-dependence argument up to this point, it doesn't really seem contentious. But it becomes so if we take it one stage further and make the following suggestion. Perhaps in the evolution of animals the generalized improvements tend to happen early (e.g. in the Ediacaran or Cambrian periods), while subsequent evolution tends to take the form of more environment-specific adaptations. If this is true, then the evolutionary process changes in the nature of the improvements/adaptations it produces as time progresses.

But is this really true? The form given above is probably too restrictive, as can be seen when we inject evolutionary novelties (Chapter 23) back into the picture. Recall the two main examples given there: the protective shell of turtles and the venom claws of centipedes. Both are useful features in a very broad range of environments. Not only that, but neither of these novel structures has ever been completely lost in the groups concerned, despite the great potential for this to happen, given that there are hundreds of species of turtle and thousands of species of centipede.

So consideration of novelties forces us to abandon too restrictive a version of the scale-dependence view of evolution and to admit that generalized improvements have taken place later than the Cambrian. But again, this simply refines the argument further so that not just environment-specificity and probability of reversal are treated as continua but so too is geological time. Thus we no longer think in the binary way of before *versus* after the end of the Cambrian, but rather in the more normal temporal way of 'how long ago'. Perhaps the further forward we come in geological time, the rarer the origins of generalized improvements?

There is much scope for refining the scale-dependence argument further in the future. But for now, I will finish with a few comments on the terms used to refer to adaptation. We already noted that the nature of the selection favouring an adaptation may change over time, in which case we talk about *exaptation*. Also, although I haven't emphasized it before now, any new feature must integrate well with the rest of the bodily structure and organization – this is referred to as *coadaptation*. But the key feature of the above argument has been that some selectively driven improvements, such as the origin of jaws, are not environment-specific; rather, they turn out to be useful under a very wide range of environmental conditions. So far I've simply called these 'improvements'. But that will hardly do, because other selectively driven adaptations that are very environment-specific, such as melanic moths, are improvements too, albeit in a more restricted context. Maybe we need some new term for those improvements that seem to be at the very general end of the spectrum, but no term readily springs to mind. Perhaps 'omni-adaptations' would work? Actually, I think not; that was just a tongue-in-cheek suggestion. Instead, bearing in mind the probabilistic nature of the argument, it's better not to introduce a new term at all, but rather to label our key continuum as being *the degree of environment-specificity* of adaptations.

There's another reason not to introduce a new term for adaptations of great environmental breadth, such as jaws. Although it turns out with the benefit of hindsight that jaws are advantageous across many environments, their evolutionary origin must have taken place under some particular set of environmental conditions. The process of natural selection that happened under those conditions was, as usual, blind to what would happen later. So, just as an initial adaptation cannot anticipate a subsequent exaptation, nor can it anticipate its own broad applicability and hence its near-irreversibility. Thus here, as ever, we need to be careful about distinguishing between evolutionary events themselves and our own understanding of them.

27 The direction of evolution

The great American palaeontologist G. G. Simpson, in discussing Darwinian and other evolutionary theories, made the following point in 1953: "The various major schools of evolutionary theory have arisen mainly from differences of opinion as to how evolution is oriented." Of course, in the Darwinian school, natural selection is seen as the main cause of evolutionary orientation, or direction. In other words, this school of thought has it that natural selection steers evolution and thus causes particular lineages to go in one direction, say increasing body size, rather than another. This differs from the earlier Lamarckian approach, which involved the inheritance of acquired characters – now known not to occur – in the determination of evolutionary direction.

Natural selection is a systematic rather than random process. But its raw material is variation, and this derives ultimately from genetic mutation, which is often said to be random. However, this latter point needs some elucidation. A key assumption of Darwinian theory – sometimes explicit, sometimes implicit – is that the variation upon which natural selection acts is random with respect to whatever would be favoured under the prevailing environmental conditions. In other words, the variation is considered not to be biased in favour of forms that would have increased fitness. This is generally believed to be correct.

But variation may be biased in other ways. In particular, random mutations at the DNA level may be transformed into non-random variations at the morphological level due to the nature of the developmental process, in which some kinds of variation seem to be, in a sense, more easily come by than others. For example, a study of the roundworm *Caenorhabditis elegans*, a species that we looked at in Chapter 10, showed that about 20% of mutations that affected body length increased it, in contrast to the other 80%, which caused a reduction. The authors of this 2002 study, Portuguese biologist Ricardo Azevedo and colleagues, noted that the bias may be even greater for body volume than for body length, because many of the mutations making the worms longer also make them thinner.

If we extend this line of thinking – about the relative ease of changing animals in different directions – more broadly, it's clear that, as well as

FIGURE 27.1 Examples of limb reduction and loss in the squamate reptiles: an amphisbaenian (left) and a skink (right). Reduction and loss of limbs is an example of an easy form of evolutionary change, and is thought to have occurred more than 50 times in the tetrapods.

there sometimes being a bias in favour of smaller versus larger body size (or the other way round), changes in size in general are easier to come by than many other sorts of change. Also, changes in the sizes of particular parts of the body relative to others are easier to make than radical rearrangements; let's contrast one example of each of these to see the point more clearly.

An example of relative size change involves the length of the limbs of tetrapod vertebrates relative to their body length. There are many cases of limb reduction, and even loss, which have taken place independently in different groups. The snakes provide the best-known case of limb loss, but there are many more, such as in the amphisbaenians and the skinks (Figure 27.1). It has been estimated that, overall, there have been more than 50 independent cases of significant reduction or total loss of limbs in tetrapods.

Now, contrast that large number – more than 50 – with the single case of the origin of a shelled vertebrate, namely turtles. Although there are many armoured vertebrates such as armadillos (a group of mammals) and placoderms (a group of extinct fish), only once has a complete, integral vertebrate shell evolved. This is presumably due to the radical rearrangements involved in making a shell (which we looked at in Chapter 23) being harder to come by, in terms of generating the relevant variation, than changes in the relative sizes of different body parts, which are comparatively easy to come by.

If it is indeed true that more lineages become limbless than become shelled due to the availability of the relevant variation, then we have a situation where the developmental process itself, and the ways in which it can be altered – either easily or at all – is involved, as well as natural selection, in determining the direction of evolution.

The area we are getting into here has been given various names. If some types of change in a developmental system are impossible or improbable, the situation has been referred to as *developmental constraint*. But a constraint away from some variants also implies a bias *for* others. In fact, the term *developmental bias* can cover both sides of the coin – a bias for the production of some variants and against the production of other ones. I have argued the case for the use of 'developmental bias' before – see my 2004 book *Biased Embryos and Evolution*. Another term that has been used in relation to these ideas is *evolvability*. Some features, such as altered body size, which can be achieved by a scaling effect, are said to be more evolvable than others that require radical rearrangement of body parts.

In extreme cases, there may be what is called *absolute constraint*. This phrase refers to situations in which the number of cases of the production of a variant is not just low but zero – in other words, evolution has never gone that way and may never do so, because the changes in development required are impossible. The dragon design that involves an extra pair of limbs in a land vertebrate seems, as noted in Chapter 1, to be an example of this. No tetrapods have ever evolved an extra pair of limbs. This could be referred to as zero evolvability of such a form, as an alternative to absolute constraint against its production – these are just different ways of saying the same thing. (Care is needed, though, because we don't know what evolution might produce in the future.)

The most famous paper to be published on this general issue of constraint/bias in the last half-century was written by the American evolutionists Stephen Jay Gould and Richard Lewontin, back in 1979. These authors relished the heretical nature of what they were saying, and took pleasure in highlighting it. They said that:

> organisms must be analysed as integrated wholes, with *Baupläne* [= body plans] so constrained by phyletic heritage, pathways of development and general architecture that *the constraints themselves become more interesting and more important in delimiting pathways of change* than the selective force that may mediate change when it occurs. (my italics.)

Unsurprisingly, this paper provoked considerable controversy in the wake of its publication – a controversy that continues to this day. There is an unresolved question about what is indeed "more interesting and more important" in determining the directions evolution has taken – biases in the production of variants (constraint/evolvability) or biases in their rates of survival and reproduction (natural selection). However, in order to see the exact nature of the unresolved question, it's necessary to be clear about who or what Gould and Lewontin were criticizing in their famous paper – and it certainly wasn't Charles Darwin.

Gould and Lewontin began their paper as follows: "An adaptationist programme has dominated evolutionary thought in England and the United States during the past 40 years. It is based on faith in the power of natural selection as an optimizing agent." Indeed, their paper was sub-titled "a critique of the adaptationist programme". With regard to the origin of this programme, they pointed out that it did not lie with Darwin, who was a pluralist in terms of evolutionary mechanisms, always ready (some would say too ready) to accept roles for other factors along with natural selection. Rather, Gould and Lewontin pointed the finger at Alfred Russel Wallace (and also at the German biologist August Weismann).

Wallace's pan-adaptationist approach is well evidenced by a quotation from an essay published in 1870:

> Universal variability – small in amount but in every direction, ever fluctuating about a mean condition until made to advance in a given direction by 'selection', natural or artificial – is the simple basis for the indefinite modification of the forms of life.

This is an incredibly strong statement. It identifies selection as the sole agent that determines the direction of evolution. It allows no role for other factors – either the structure of the variation upon which selection acts (our focus here) or any others. This statement could be taken as the central tenet of what Gould and Lewontin called the adaptationist programme.

It's also important to enquire how the adaptationist programme that Gould and Lewontin criticize can be mapped to later evolutionary thinking, both in the period to which they refer – the 1930s to the 1970s – and also extending right up to the present day. Darwinian thinking from about 1930 onwards is often referred to as neo-Darwinism – that is, a version of Darwinism that has been refined by the introduction of both Mendelian genetics and a mathematical approach to the way that selection works in populations. Neo-Darwinism in turn is often associated with the disciplines of population and quantitative genetics. However, one quantitative geneticist, the American James Cheverud, responded to the Gould–Lewontin paper by saying that quantitative genetic theory already dealt with the main thing that Gould and Lewontin were championing – developmental constraint. So, in a sense, he was asking what all the fuss was about. The answer to that question is that many evolutionists had, in the run-up to 1979, taken a less enlightened view than Cheverud. Instead of being neo-Darwinians, they could be described as neo-Wallaceans. It was these that Gould and Lewontin criticized.

Interestingly, Gould and Lewontin did not refer to any of the 1930s pioneers of population genetics. They looked back before them, to Darwin and Wallace in the nineteenth century; also, they criticized what was then recent work – in particular the work of several evolutionists in the 1970s. This lack of references to the 1930s was a serious omission, because one of the founders of population genetics, the British evolutionist J. B. S. Haldane, had in a way pre-empted their point. Here is a quotation from Haldane's 1932 book, *The Causes of Evolution* (note the pluralism embodied in that title):

> To sum up, it would seem that natural selection is the main cause of evolutionary change in species as a whole. But the actual steps by which individuals come to differ from their parents are due to causes other than selection, and in consequence evolution can only follow certain paths. These paths are determined by factors which we can only very dimly conjecture. Only a thorough-going study of variation will lighten our darkness.

This shows a pluralistic approach, similar to Darwin's. Also, it acknowledges that biases both in the production of variants and in their survival may be important, rather than there being an either/or situation. And it's helpful in taking the positive approach of focusing on variation rather than constraint, albeit the two are inextricably linked. If variation is easy to produce in some directions but hard or impossible to produce in others, then this state of affairs may push evolution into certain channels and constrain it from going in other directions.

It's time for an example. As noted earlier, there are several thousand species of both mammals and birds. In both of these groups, there is considerable variation in body size and body shape. Let's focus in particular on neck length, which varies partly in association with overall body size, but partly also independently from it. In mammals the obvious example of a long neck is the giraffe. In birds there are competing examples: ostriches, swans and flamingos spring to mind, along with several others. Comparing a giraffe with a human or other ordinary mammal (ordinary in terms of neck length, that is), it's interesting to note that the number of cervical (= neck) vertebrae is the same: seven. In fact, almost all species of mammals have a fixed number of seven neck verte-brae; the only exceptions are some species of sirenians (sea cows) and of sloths, groups that are not exceptional at all in terms of their neck lengths.

In contrast to the almost complete lack of variation in the number of mammalian neck vertebrae, there is considerable variation in this number in birds. Often, those birds with long necks have more cervical vertebrae than those with short ones. For example, swans have between 20 and 25 neck vertebrae, while the short-necked parrots typically have from 10 to 12 (Figure 27.2).

There are two ways to interpret this striking contrast between mammals and birds. One is that the mammalian and avian developmen-tal systems are different in some way that we don't yet understand – and that this difference means that it is easy for variants in cervical vertebra number to arise in birds but much harder for them to be produced in mammals. The other is that when selection favours long necks in mammals it only does so in those cases where the lengthened necks are produced by elongating the standard seven vertebrae; and it acts against variants in which the elongate necks are based on a greater vertebral number. Neither hypothesis can be excluded at present. However, a claim that the second one is right would fall squarely into Gould and Lewontin's adaptationist programme.

There is an interesting twist to this story. Suppose that variants with more or fewer than seven neck vertebrae can be produced by mamma-lian developmental systems, but that they are detrimental or even lethal during embryogenesis. In this case, mammals that are born will show little or no variation in their number of cervical vertebrae, but mammals in the womb will show much more. Do we have any evidence for or against this possibility?

Strangely, we have both. The American palaeontologist Robert Asher and his colleagues analysed data on several species of mammals in 2011 and found that embryos did not have more abnormal vertebral numbers than adults of the same species. However, the Dutch biologist Frietson

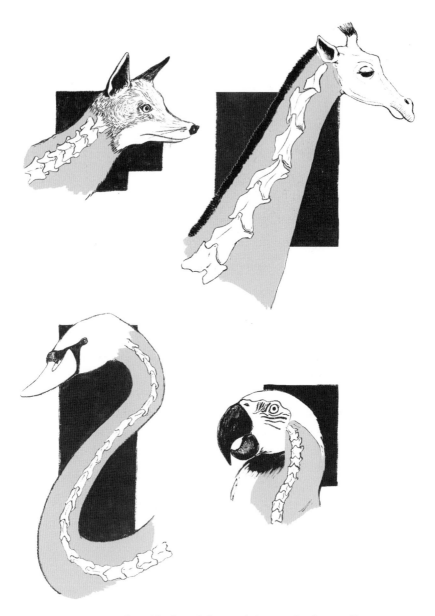

FIGURE 27.2 Mammals and birds with long and short necks: fox, giraffe, swan, parrot. The neck vertebrae are shown in each case. As can be seen from comparison of these numbers, long-necked mammals evolve by elongating their seven cervical vertebrae, while long-necked birds evolve by increasing their vertebral number (in the comparison shown the numbers are 11 and 22). The reason for this difference between birds and mammals is not yet clear.

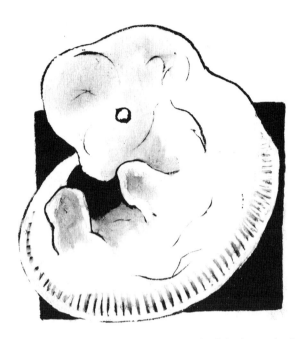

FIGURE 27.3 A mouse embryo at a stage when the limb-buds are clearly defined. These buds constitute developmental modules. The various internal organ primordia can also be thought of as modules in the same way.

Galis and her colleagues, studying humans, showed, in a paper published in 2006, that variant numbers of cervical vertebrae are higher among embryos that died than among those that survived, as represented by the population at large. This observation suggests that part of the reason for the constancy in the number of human neck vertebrae is a form of selection – but not selection that is related to adaptation to the external environment because most of this selection is going on *in utero* (though further selection may occur later in life also).

The British polymath Lancelot Law Whyte called this kind of process *internal selection* in his 1965 book *Internal Factors in Evolution*. Perhaps this was an unfortunate choice of phrase. Maybe 'selection for organismic integration' would be a better one. Or selection for coadaptation. Whatever we call it, this process is interesting because it blurs the distinction between developmental bias and natural selection.

There are two other things that need to be introduced into the debate. The first of these is modularity. Limb-buds are often thought of as modules whose development is quasi-independent of other parts of a developing embryo (Figure 27.3). Organ primordia, such as those that give rise to eyes or lungs, can also be thought of in this way. Most if not all animal embryos are modular to some extent. However, the extent varies with developmental time. Modularity may be lowest during the phylotypic stage (or period) that we examined in Chapter 17. This can be equated with that stage being the most constrained – or the least evolvable.

The other thing we need to look at here is the size of step by which evolution proceeds, because, if step size is large, mutation may be a determinant of evolutionary direction. This is a topic on which Darwin was uncharacteristically dogmatic, a topic in relation to which he was not his normal pluralist self. He made use of the Latin diktat *Natura non facit saltum* (nature does not make leaps); interestingly, he was criticized by his otherwise staunch supporter T. H. Huxley for doing so. Darwin's vision – in this respect very similar to Wallace's – was of evolution taking place by a gradual accumulation of very small changes.

Over the years, there have been various rebellions by evolutionists against this gradualist approach. Probably the most famous such rebel was the German-American geneticist Richard Goldschmidt, whose views are usually described as saltationist – and so in direct conflict with Darwin's gradualism. Goldschmidt was a student of those big-effect mutations that we examined in Chapter 19, the homeotic mutants in which the right structure comes to be in the wrong place – such as an extra pair of wings in flies developing on the third thoracic segment where there should normally not be wings but instead small flight-balancing organs. Goldschmidt felt that these mutations showed how major evolutionary changes took place – all at once instead of by accumulation of lots of tiny modifications.

In a paper published in 1952, entitled "Homeotic mutants and evolution", Goldschmidt says that "real evolutionary changes ... are brought about by large mutational steps, saltations, which change at once the major features of early development." In case this is not clear enough, he goes on to deny that important evolutionary changes can be produced other than by a leap or saltation: "all major evolutionary steps involving real change of organization cannot be produced in any other way than by an initial change of determination in early development."

Goldschmidt's saltational theory of evolution has been rejected by most evolutionary biologists. The main reason for this is that homeotically mutant *Drosophila* fruit-flies (the animals in which this phenomenon has been most studied) are known to be very unfit compared to their non-mutant counterparts – a decrease in fitness that is general rather than environment-specific. Thus they would be eliminated from, not spread through, natural populations by Darwinian selection. However, before dismissing Goldschmidt entirely, two observations need to be made about him and his theories.

First, it is interesting to note that many proponents of the new evo-devo approach to evolution hold Goldschmidt in high regard. For example, the American biologists Rudolf Raff and Thomas Kaufman dedicated their 1983 book *Embryos, Genes and Evolution* to him. This should not be taken as indicating support for Goldschmidt's version of saltationism though. Rather, it's because in the 1950s mainstream evolutionary theory – the modern synthesis, as it was called – excluded development and concentrated too much on populations and not enough on individual organisms. Both are important in evolutionary theory. Goldschmidt emphasized this point at the end of his 1952 paper, where he says, in an exasperated tone: "I have tried for a long time to convince evolutionists that evolution is not only a statistical genetical problem but also one of the developmental potentialities of the organism."

Second, while evolutionary saltation via homeotic mutation in flies may never have happened, this does not mean that large-effect mutations in general have never contributed to the evolution of any group of animals. They have. Two examples serve to illustrate this point. Snail-shells can reverse their direction of coiling from right-handed to left-handed by a single mutation, as we saw earlier (in Chapter 9). This kind of reversal seems to have happened many times in the evolution of the gastropods. The other example involves centipedes, and particularly those belonging to the group that contains the large, dangerous tropical species, namely the group that is called Scolopendromorpha.

There are about 700 species of centipedes in this group. Up to the year 2008, it was thought that all of these had either 21 or 23 pairs of legs. However, the discovery of a previously unknown species of scolopendromorph in that year by the Brazilian biologist Amazonas Chagas-Junior changed the picture entirely. This species has approximately double the number of pairs of legs possessed by all other species of its group – a fact that is reflected in its name: *Scolopendropsis duplicata*. The most plausible explanation of its origin is a large-effect mutation that all-at-once led to a doubling of the number of trunk segments, and thus of the number of leg-pairs.

It may be possible to generalize from these two examples – snail chirality and centipede segment number – as follows. The idea that the developmental effects of mutations can be arranged in a simple linear sequence that can be thought of as generalized magnitude of effect is wrong. Even more wrong is the drawing of a line at some point in such a linear sequence above which are macromutations and below which are micromutations. The latter approach is undesirable because there is no obvious point at which to draw the line. The former is undesirable because it takes no account of different *types* of change in development. Even those who have been highly critical of saltationist theories of evolution have admitted this point. For example, Richard Dawkins goes saltationist-bashing in his 1986 book *The Blind Watchmaker*. He dismisses macromutations as possible bases of evolution, but with an important proviso: he admits that what he calls "stretched DC8" macromutations may indeed have contributed to evolution.

Dawkins uses this term to refer to the production of a longer version of the same design of aircraft, in contrast to the sort of macromutation that could, for example, all at once turn a helicopter into a plane – or even assemble a plane from scratch. (An aside here: the Douglas DC-8 was an airliner first built in the late 1950s, with longer, 'stretched' versions appearing in the late 1960s). No doubt Dawkins would have placed snail chirality mutations into his "stretched DC8" category, because these mutations were known at the time. However, ironically, we now have something that is even more obviously a "stretched DC8" – namely our stretched centipede with its duplicated trunk.

I would argue that the way forward is to probe further into the structure of variation, as recommended by Haldane a long time ago. Of course, there have been many studies of variation post-Haldane; but there is much further work needed. We are still to a large extent in the darkness of ignorance to which Haldane referred. And future studies should not concentrate exclusively on so-called micromutational effects. Indeed, the characterizing of variation in terms of type rather than merely size of effect is a major priority. So too is establishing the *timing* of effects in terms of which stage of development they first begin to alter. If I was asked what I think will be the most important area of study for the advancement of evolutionary theory over the next couple of decades, this is the one I would identify.

28 Animal extremophiles

Although the subject matter of this chapter is very different from that of the last one and the next one, a common theme running through all three is probability, or perhaps – though it's really the same thing looked at from a different perspective – improbability. In the last chapter we saw that the nature of the developmental system made the evolution of certain kinds of animal improbable. In the next chapter we look at the probability of animal-type life on other planets. Here, we look at animals called extremophiles that live in parts of our own planet where animal life would at first sight seem improbable, and/or that have tolerance of extreme conditions.

Although these animals are indeed called extremophiles, this term is used also to describe other forms of life, from other kingdoms, that can withstand extreme environments. So the term can be used also for plants and fungi, and especially for organisms from the bacterial and archaean domains, where an extremophile existence is most commonly found. But here we'll concentrate on extremophile animals. We'll also concentrate on extremes of temperature; but it's worth noting that extremophile can be used in relation to other environmental variables – for example acidity.

Extremophile animals are best approached by contrasting them with ordinary animals – in the sense of those that inhabit non-extreme environments and are unable to tolerate exposure to extremes of various kinds. Humans are an example of such ordinary animals. We're an unusual example, though, because of our ability to alter our immediate surroundings so that the conditions we experience can be very different from those prevailing a short distance away. I am writing this chapter in an ambient temperature of about 20 degrees Celsius; yet two metres away the temperature is only about 3 degrees Celsius.

Many non-human animals can also modify their immediate surroundings, though their ability to do so is less pronounced than ours. Turtles, as we saw in Chapter 25, can excavate a burrow or nest in which to lay eggs, thus reducing the variation in temperature to which the developing embryos are exposed. This is important, given the temperature-dependent sex determination mechanism in these animals. There are many other examples that could be given of animals creating their own ambient temperatures – and they are not all vertebrate ones. The temperature within a honeybee nest or a termite mound can also be quite different from that which applies outside.

Most animals do not build elaborate nests and are at the mercy of whatever temperature prevails in the surrounding environment. However, even then, they can alter the conditions they experience through selecting a particular microhabitat characterized by its own microclimate. In Mediterranean regions where the daytime temperature in summer is dangerously hot for small invertebrates, land snails can be seen aestivating. This is the summer equivalent of hibernating in winter. The snails climb to the top of long stalks of vegetation, because it is significantly cooler there than close to the ground. Then each snail secretes a membrane that simultaneously covers the aperture of its shell and attaches it to the stalk it has climbed up.

What I've called ordinary animals above (i.e. non-extremophiles) live at a range of temperatures. The average temperature they experience during their lives varies from just above freezing to about the internal temperature of the human body – 37 degrees Celsius. Of course, an average hides much variation. Both the amount of variation and the average depend on many things, including latitude. Alaska is not only much colder on average than central Africa; it also has a much wider range of temperature variation between seasons.

Extremophile animals can live outside this range of temperatures. Some live almost all the time in sub-zero conditions; others thrive at temperatures in excess of 50 degrees. A few animals can withstand really extreme temperatures, such as minus 200 degrees and beyond, as we'll shortly see. However, note that an important difference has just crept into the story. I've just said that some animals live in or thrive at extreme temperatures; but I've also said that some can *withstand* extremes. These two things are not the same. Strictly speaking, *extremophile* should only refer to the former, because the word means seeking extremes – in other words actively preferring them or even requiring them. However, I'm using the term in a broader sense here, to include animals that can tolerate extreme conditions even if they don't normally experience them.

Let's look at three animal extremophiles: one each from the three major branches of the Bilateria. We'll look first at icefishes

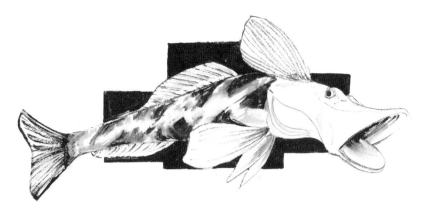

FIGURE 28.1 A crocodile icefish. These fish are adapted to life in unusually cold water – between zero and minus 2 degrees Celsius, the latter being the approximate temperature at which seawater freezes.

(Deuterostomia); then at water-bears (Ecdysozoa); and finally at a strange group of segmented worms (Lophotrochozoa).

There are several families of Antarctic icefishes. We'll concentrate here on just one of these – the so-called crocodile icefish (Channichthyidae). Their alternative name, white-blooded icefish, derives from the fact that their blood lacks haemoglobin. Overall, there are about 30 species within the family, one of which is shown in Figure 28.1. They mostly live in the cold water around and underneath ice-sheets. The temperature experienced by the most southern species is typically within a rather narrow range between about minus 2 degrees and zero. (Temperatures of below minus 2 are not experienced by these fish because that is the approximate temperature at which seawater freezes.)

The crocodile icefish are able to survive without haemoglobin (and indeed without the red blood cells that other vertebrates carry their haemoglobin in) by transporting oxygen in the blood plasma. They also have unusually large blood vessels and an extremely low metabolic rate. Crucially, their blood and other bodily fluids contain antifreeze proteins that attach to the surface of any tiny ice crystals that form and stop these crystals growing. Without this particular adaptation the icefish would die, because their blood is more dilute than seawater and hence it would freeze at approximately minus 1 degree. The antifreeze proteins reduce the freezing point of the blood from about minus 1 to minus 2 – a small but critical amount. Further details can be found in the 2002 book *Life at the Limits* by David Wharton.

FIGURE 28.2 A tardigrade or water-bear. These animals are tiny – often less than 1 millimetre in length. They are closely related to the arthropods. They are famous for their ability to survive extreme conditions, including a temperature only 1 degree above absolute zero.

The crocodile icefish are not primitive creatures that represent some early branch of fishes. Rather, they belong to the main group of modern fishes – the teleosts. And, in general, animal extremophiles are not ancestral but instead are derived from ordinary forms by specialization to extreme conditions. There are important implications of this fact for theories about the possible evolution of life on other planets or moons (see next chapter).

We'll now turn to water-bears or, to give them their official name, tardigrades. These tiny creatures, most of them less than a millimetre long, are closely related to arthropods, as we saw in Chapter 14. The appropriateness of their common name can be seen from Figure 28.2: they have a distinctly bear-like appearance, despite their diminutive size. Their official name means slow-mover or slow-walker – an indication of their ponderous gait, something that we would need a video rather than a figure to observe. Tardigrada is in fact a phylum with more than 1000 described species.

The habitats occupied by extant tardigrades are very varied – including terrestrial, freshwater and marine ones. Terrestrial tardigrades often live in moss or lichen. They can be found at very high densities – sampling just a few cubic centimetres of moss from an old wall can yield hundreds of them. Not all tardigrades live in extreme conditions, such as hot springs or on deep ocean floors. Even the conditions experienced by the species inhabiting such environments would not explain tardigrades' incredible ability to withstand extremes of temperature. In laboratory experiments, tardigrades have been subjected to temperatures as high as 150 degrees Celsius and as low as minus 272, which is only 1 degree above absolute zero; in both cases they have survived.

Although we can't answer the 'why?' question about tardigrades' ability to withstand such extreme conditions, we do know something about the 'how?'. When their surroundings become such that life is barely possible, including extreme temperatures and extreme desiccation, tardigrades can undergo a transition into a strange form called a tun. The legs are retracted, most body water is lost, and the metabolic rate drops so low that it is virtually zero. The creature is capable of remaining in this state of suspended animation (or cryptobiosis) for long periods of time; it is also capable of returning to normal when conditions become more favourable again. Although most of our current understanding of the transitions to and from cryptobiosis is at the level of gross morphology and physiology, studies of the molecular basis of these transitions have now begun, and they promise to yield some very interesting results.

Tardigrades have also been taken into space to test their ability to tolerate extreme conditions. And 'taken into space' should be interpreted literally. Unlike some of the more famous space-travelling animals, such as monkeys, dogs and mice, tardigrades were exposed to the vacuum of space (on the 2007 Foton-M3 mission) rather than being protected from it. Most of them survived the experience.

Tardigrade fossils are known from the Cambrian period – more than 500 million years ago. So the phylum has a long history, with rather little change in its constituent creatures over time. However, just as the icefish represent an internal clade within the vertebrates, the tardigrades are a clade within the Ecdysozoa; they are not a basally

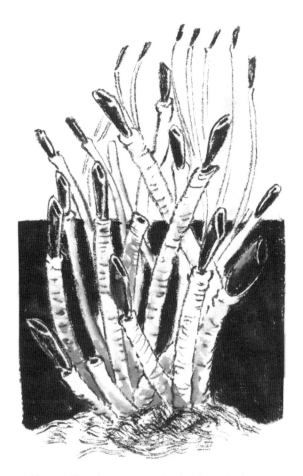

FIGURE 28.3 A clump of beard-worms, or siboglinids, with their posterior ends attached to a marine substratum and their anterior ends projecting up into the seawater.

branching phylum within this super-group but rather are within the Panarthropoda – the group mentioned in Chapter 14 that includes the true arthropods, the tardigrades and the velvet worms. Again, this has implications for the origin of extraterrestrial life, and in particular the question of whether it might be able to originate in extreme environments.

Our third group of extremophile animals is a family of segmented worms called the beard-worms or, more officially, the Siboglinidae (Figure 28.3). These animals live in tubes at various ocean depths. Some species are known only from great depths, and indeed from just

FIGURE 28.4 A black smoker – a type of hydrothermal vent found at great depths on the ocean floor, often at mid-ocean ridges. Some beard-worms can be found attached to these structures, despite the very high temperature.

one very specific type of habitat – the vicinity of black smokers. These chimney-like structures (Figure 28.4) are found at various points along mid-ocean ridges, which are examples of a particular type of plate boundary in the Earth's crust – a divergent boundary where material is upwelling and the plates thus being formed are slowly moving away from each other.

Beard-worms can be found actually attached to the sides of the smokers. This location means that they are subject both to very high temperatures and to amazing temperature variation. The water emerging from black smokers is superheated; it's at about 300 degrees Celsius, something that is possible, without the water turning to steam, because of the very high pressures prevailing at these great ocean depths (a few thousand metres). However, not far away, the deep ocean temperature is only a few degrees above zero. It has been estimated that one end of a worm living in a tube that is attached to a black smoker may be experiencing a temperature of above 80 degrees Celsius while the other end might be at a mere 20 degrees. The exact temperature difference depends on the length of the worm; and these animals, unlike tardigrades, can be rather long. Depending on the species, their maximum length can be anything up to 1.5 metres.

These worms form part of an unusual kind of ecosystem – one that is powered by chemosynthesis instead of the usual photosynthesis. The energy-fixing organisms in these chemosynthetic systems are bacteria (and archaeans) rather than plants. They obtain energy from sulphides emanating from the black smokers. Some of these bacteria live in the tissues of the tube-worms, with the latter obtaining their energy from the former. Since the worm is providing the bacteria with a home, this could be said to be a symbiotic or mutualistic relationship, in which both species gain from the presence of the other.

When looked at from an evolutionary perspective, the message is the same as for icefish and tardigrades – the beard-worms constitute a clade within the phylum Annelida (segmented worms) rather than being a basally branching group within the super-group Lophotrochozoa. However, that conclusion was only reached quite recently. When I was a student learning about the different animal phyla, the worms that are now placed in the annelid family Siboglinidae were thought to belong in not just one but two unique phyla, of which they were the only members.

As we've seen, all three groups of extremophile animals that we've examined have been nested within larger groups, the majority of whose members are not extreme at all. So, in each case, an extremophile ability or ecology has arisen from something much more ordinary. The larger group in each case probably originated in non-extreme conditions. Indeed, this may be a special instance of the pattern noted in Chapter 26, with specialized adaptations arising after, and on the basis of, broader ones.

To what extent can this pattern be extended to the origin of the animal kingdom? If most or all animal phyla arose in 'ordinary' conditions, might the same be true of animals overall? This is a tricky question to answer, given that a key feature of the origin of animals was the advent of multicellularity, which, in the context of the animal kingdom, is a unique, non-replicated event.

At least we can be reasonably confident of the group of unicells from which the animal kingdom stemmed: the choanoflagellates (or collar-whips) that we looked at in Chapter 2. There are about 200 species of these alive today, and most are not extremophiles. Not enough is yet known about the evolutionary relationships within the group for it to be apparent whether the few species that inhabit extreme environments, notably waters under Antarctic ice-sheets, are basally branching or not. And the really interesting question of what the original (pre-Cambrian) choanoflagellates were like and what sorts of environment they were found in is unanswerable because the group has virtually no fossil record. I'd hazard a guess that the origin of colonial choanoflagellates, and hence the origin of animals, took place in shallow marine waters where conditions were not extreme. However, that is indeed a guess – or you could dignify it with the term hypothesis if you like; it's not an established fact.

The origin of life, though, is a different matter altogether from the origin of animals. The gap in time between the two events was probably more than 3 billion years, though, as we saw at the start of the book, there is considerable uncertainty about the timing of both. When life began, around 4 billion years ago (BYA), at the end of the interestingly named Hadean aeon when the Earth was still cooling down from its fireball beginnings, most parts of the Earth's surface were probably extreme, at least by our current standards. So perhaps life can originate in extreme conditions, and perhaps animals cannot. Again, though, these are merely hypotheses.

A final hypothesis is that intelligent life generally does not evolve under extreme conditions. Certainly that was the case here on Earth, in terms of the origin of humans, which we examined in Chapter 24. No species of intelligent life-form has evolved here on Earth under extreme conditions of temperature or other environmental variables.

Is there a message for the evolution of intelligent life on other planets? Probably. But then again, what's extreme for Earth-based life may not be so for some entirely different array of life-forms found elsewhere, especially if they are not carbon-based. The following chapter will take a look at the possible evolution of life in general, and intelligent life in particular, on other planets.

29 Extraterrestrial animals?

Are we are alone in the Universe? Almost every adult human alive today has asked this question. About 400 years ago, when Galileo watched the four large moons of Jupiter orbit the giant planet, he probably asked himself that very same question. Perhaps even when *Homo sapiens* first arose in Africa some 200,000 years ago, our ancestors looked up at the night sky and wondered if there might be life 'out there' somewhere – albeit their notion of the Earth and what lay beyond it must have been very different from ours.

However, the question 'are we alone?' needs some dissection. The answer may depend on what is meant by 'we'. Does the 'we' of the question refer to life, animal life, or intelligent life? The probability of us being alone must increase as we move through this series of gradually more restrictive categories. And yet, it may be that for all versions of 'we' the answer is no – we are not alone.

Before looking at the possibility of life having evolved elsewhere than on Earth, let's probe a bit into each of the three kinds of 'we', as listed above, starting with simply life. It's not easy to define life. In an Earth context, the main properties that life-forms have, which non-living things such as stones do not, are reproduction and inherited variation. These are the very things that are needed for Darwinian natural selection to operate. Selection can enable bacteria to become better adapted to their environment, but it cannot do the same for stones. Whether using these same features to distinguish life from non-life in an extraterrestrial context is sensible remains an open question. However, a reasonable strategy is to use them for now but to keep in mind that changing to other criteria might later prove to be necessary.

Notice that this approach to defining life does not use any criteria that relate to what it is made of. On Earth, all life is carbon-based; most life is cellular; and most life uses DNA as its hereditary material. But extraterrestrial life might not have any of these features.

The assumption that any life on any planet anywhere in the Universe would have to be carbon-based has been referred to as carbon chauvinism. I would advocate a dual approach to this issue. First, I think that we would be wise to acknowledge that there are other chemical elements, notably silicon, which can form the basis of large complex molecules, and so, possibly, life. But second, I think we should bias our strategy in searching for extraterrestrial life in favour of looking for the conditions in which carbon-based life could thrive. After all, we *know* that carbon-based life is possible; we can only speculate that silicon-based, or 'other-based', life is also possible.

Assuming that all life on other planets would have to be cellular and to use DNA as its hereditary material would be foolish, given that there are organisms on Earth that are lacking in one or both of these features. The syncytial slime moulds attain macroscopic size despite the fact that their bodies are not divided into cells; and some viruses have RNA, not DNA, as the macromolecule used to store their genetic information (Figure 29.1).

The second usage of 'we' was in the sense of 'we animals'. It's entirely possible that there are planets on which many rock surfaces are covered in green slime, and which are thus teeming with life, yet not a single animal has evolved. Recall, though, from Chapter 1, that not only is it hard to define life, it is also hard to define animals. Here on Earth the safest way to do so is to use 'animal' only for members of the great group of multicellular life-forms that arose from a single stem among the unicellular choanoflagellates. But using this definition, no extraterrestrial life-forms could be called animals. So instead we should use the alternative approach and say that animals have the following features: multicellularity; heterotrophy (i.e. eating organic material rather than photosynthesizing); and active movement of at least one stage of the life-cycle. Recall that the last of these serves to separate animals from fungi, which are heterotrophic but have only passively mobile spores.

Finally, let's consider the third 'we': intelligent life-forms. How should intelligence be defined? In Chapter 12, when dealing with the issue of octopus intelligence, I took an approach of inferring intelligence from certain types of behaviour. In particular: learned rather than instinctive behaviour; flexible as opposed to fixed behaviour; and types of behaviour that seemed to imply planning of some sort. These were fine for our purposes then. But they need to be added to for our present purpose. This is because human interest in intelligent extraterrestrial life is usually focused on life-forms that are more like us than octopuses are. Specifically, by 'intelligent extraterrestrial life-forms', we usually

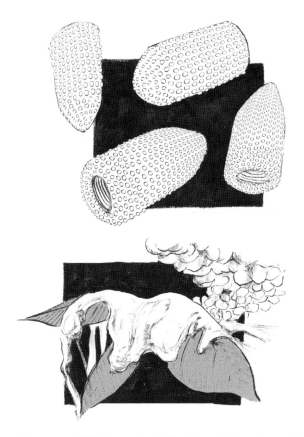

FIGURE 29.1 Two unusual life-forms: the bullet-shaped rabies virus, which is an example of an RNA virus (top); and a syncytial slime mould, growing over some leaves (bottom). The former has its genetic material encoded in the form of RNA as opposed to DNA. The latter is large enough to be a multicellular creature but its body is not divided into cells; rather, many nuclei are found in a large undivided cytoplasm.

mean creatures that have developed an advanced technology, as we humans have. One pragmatic choice for defining a technology as being advanced is the ability to send and receive radio signals, since this is how we are most likely to discover (and be discovered by) intelligent life on other planets.

The use of 'planets' here merits some discussion. First, given that there may be forms of extraterrestrial life that are very different to our preconceptions of what life should be like, it's not impossible that there is life in the middle of a star (i.e. a sun) or in the near-vacuum of interstellar space. I'm not going to discuss these possibilities further because personally I don't believe that such forms of life exist – though of course I could be wrong. Second, 'planets' should be taken to include the moons orbiting planets; there has already been considerable speculation about whether there might be life on some of the moons of Jupiter or Saturn – more on this below. Third, 'planets' is sometimes taken to refer only to the eight planets belonging to our own solar system. But the discovery of planets orbiting stars other than our Sun – exoplanets – has mushroomed over the last 20 years or so. In fact, exoplanets may be the most likely places to harbour extraterrestrial intelligence.

It's now time to consider the possibility of life, animal life, and intelligent life in various extraterrestrial places. Let's start with our own solar system. By now, after sending out multiple space-probes, including many landers, and having broadcast radio waves into space that would have reached all parts of the solar system decades ago, it seems clear that there is no intelligent life on any planet or moon belonging to this system. There is probably no animal life either, though that's harder to be sure about because a shrimp-like creature living in an underground sea on one of the moons of Saturn would be invisible to a passing spacecraft and would be unable to perceive or respond to our radio signals.

What about the remaining possibility – that there is life that is neither intelligent nor animal – somewhere in our solar system? Might there be bacterial cells, or something like them, on Mars or elsewhere? It's probably sensible to approach this question in a systematic way, starting from the planet closest to the Sun – Mercury – and working our way outwards. To do this, it will help to refer to Figure 29.2, which provides a reminder of the eight planets of our system and their order of occurrence.

Mercury and Venus – the two planets closer to the Sun than we are – can be easily disposed of, at least with regard to carbon-based life as we know it. Mercury has virtually no atmosphere and it is a planet of extreme temperatures – mostly but not entirely in the hot direction. Extreme cold can also be found on Mercury because it has only a very slight tilt to its axis; the polar regions therefore get little heat – it's thought that ice is present there at the bottom of deep craters. It is also interesting to note that Mercury orbits the Sun very quickly – in about three Earth months – but rotates on its axis only three times for every two orbits (a ratio that is maintained by a process called gravitational resonance). This means that a Mercurian day lasts for two-thirds of a Mercurian year.

Mars   Jupiter   Saturn

FIGURE 29.2 The eight planets of our solar system. There has been speculation that some of the moons of Jupiter or Saturn might harbour simple life-forms, perhaps in subterranean oceans, which is why these two planets are named in the picture, along with Mars, which is still suspected by some scientists of concealing either current life or evidence of past life. (Note: distances and size differences have been reduced.)

Venus has been described as having a hellish surface environment. Volcanic activity is rampant. The atmospheric pressure is nearly 100 times that on Earth; and the atmosphere is made up largely of carbon dioxide. There are clouds of sulphuric acid. The typical surface temperature is around 500 degrees Celsius. This combination means that carbon-based life would have no chance of survival. The conditions survived by extremophile organisms on Earth, such as the 80 degrees experienced by beard-worms attached to black smokers on the deep ocean floor, seem mild by comparison with conditions on Venus.

Just now, as I am writing this chapter, evidence of past or present life on Mars – Earth's neighbour in the outward direction – is being searched for by NASA's robotic vehicle named *Curiosity* (Figure 29.3). This roving car-sized vehicle has tested the Martian surface for signs of life in many ways, including drilling into rocks. So far it has found no direct evidence of life. I wonder if that will still be true when you are reading this. I would guess so.

The gas-giant planets of Jupiter and Saturn would at first seem unlikely places to find life. They are much further from the Sun than is Earth or Mars, so their surface temperatures are very low; and the fact

FIGURE 29.3 The vehicle known as the *Curiosity* rover on Mars. At the time of writing, this rover has not discovered any direct evidence of past or present life. Is that still true at the time of reading?

that they are indeed largely made of gas does not bode well for making them potential homes for carbon-based life. However, their moons have attracted the attention of scientists who are interested in the possibility that extraterrestrial life exists in our solar system.

In particular, interest has focused on Jupiter's moon Europa and Saturn's moon Enceladus. Both are thought to have liquid water existing in subterranean seas. At first it seems strange to propose that water might exist in liquid form so far from the Sun on a moon whose surface temperature is a long way below zero – about minus 200 degrees Celsius. However, it is thought that some non-solar form of heat – geothermal, very broadly defined – might produce above-zero temperatures in the subterranean seas despite the intense cold of the surface. Enceladus is also interesting because water jets, presumably emanating from such subterranean seas, can be seen shooting out a considerable distance into space. These consist of tiny ice particles, because of course in space the temperature is too low for the jets to remain as liquid water.

However, despite such promising signs of water, there is as yet no evidence for life. Perhaps this is asking a bit much, because although both moons have been subject to close fly-pasts (as close as 50 kilometres in the case of Enceladus), neither has been landed on by any spacecraft. But the one moon of Saturn that *has* been landed on (Titan; by the *Huygens* lander: see Chapter 24) did not provide us with any evidence of life either. And there is a problem for enthusiasts about the possibility of finding life on any of the moons of Jupiter or Saturn. Such folk often point to the extremophiles of Earth as a positive sign. The rationale is that if extremophiles can exist here, why not elsewhere too? However, a counter-argument was seen in the previous chapter: extremophile animals here tend not to be primitive; rather they are derived from non-extremophile ancestors.

Further out in the solar system, the planets Uranus and Neptune, their moons, the dwarf planet Pluto, and other solid bodies such as Sedna all look unlikely places for carbon-based life to exist, because of the extremely low temperatures prevailing so far from the Sun. So we will not consider them here. Instead, it's time to turn our attention to the explosive growth in the discovery of planets orbiting suns other than our own – exoplanets.

In order to appreciate the difference between exoplanets and Solar planets we need to have some idea of the difference between the spatial scales involved. The distance of the Earth from the Sun is about 150 million kilometres – this is referred to as 1 AU (astronomical unit). The distance of Jupiter from the Sun is about 5 AU (750 million kilometres). If these distances are transformed into light years, they are tiny. Light travels from the Sun to the Earth in about 8 minutes. So the distance of 150 million kilometres, or 1 AU, is equivalent to a distance of about 8 light minutes, which is less than one ten-thousandth of a light year, or < 0.00001 light years if you prefer.

The nearest star system to Earth is the star trio made up of the binary star Alpha Centauri and the red dwarf Proxima Centauri. These are between 4 and 5 light years away. The north star, Polaris, is between 400 and 500 light years away – in other words a hundred times further than Alpha Centauri. And the most distant stars from us but still in our own Milky Way galaxy are many thousands of light years away. In total, the Milky Way contains at least 200 billion stars. It's now thought that most of these have one or more exoplanets orbiting them.

Let's do some rough calculations on the likelihood of intelligent extraterrestrial life. In doing these we are following in the footsteps of the many cosmologists and other scientists who have been associated with the SETI (Search for Extraterrestrial Intelligence) project, which started about 1960. In particular, we are following in the footsteps of the American astronomer Frank Drake, who came up with an equation that bears his name, the purpose of which was to show that the number of civilizations in the Milky Way galaxy could be predicted in terms of a series of fractions (or probabilities), such as the fraction of stars that have planets.

If only half of the stars in the Milky Way have a single planet and just a few have two or more, then there are at least 100 billion exoplanets in our galaxy. But the number of galaxies in the observable Universe – which was unknown in Drake's time – is now thought to be of the same order of magnitude as the probable number of stars in our galaxy – at least 200 billion. If we make the assumption that our galaxy is an average one in terms of its number of constituent stars and their typical number of planets, then the overall number of planets in the Universe is 200 billion times 100 billion. That's 20,000,000,000 trillion. The probability of none of them having life – even of none having intelligent life – must thus be very close to zero. Or, put another way, intelligent extraterrestrials with civilizations at least as advanced as our own are almost certain to be out there somewhere – and probably in many different places. Thus the kind of intelligent extraterrestrial beloved by science fiction is almost certainly real, even if it does not look exactly as in Figure 29.4.

The exoplanets that have been discovered so far are very varied in their size, their distance from their sun, their orbital period, and other features. Planets ranging from about Earth-sized to more than 10 times the size of Jupiter are known. Detection of small exoplanets is harder than detection of large ones, so planets much smaller than Earth may well be common too. In terms of proximity to their suns, there are exoplanets a mere couple of million kilometres out; but also exoplanets that are hundreds of times more distant from their suns than Earth is from our Sun. Orbital periods range from less than an Earth day to more than 1000 Earth years. Anyone interested in the fascinating diversity of exoplanets would do well to read the 2012 book *Destiny or Chance Revisited: Planets and Their Place in the Cosmos*, by Stuart Ross Taylor. And they would do well to take heed of a cautionary message therein: that thousands of exoplanets have now been discovered (with varying degrees of certainty) but that any attempt to list them would be "out of date before it appears in print".

An interesting term that has appeared in relation to the probability of life existing on exoplanets is the Goldilocks Zone. This is named, of course, after Goldilocks and the three bears; more specifically, it is

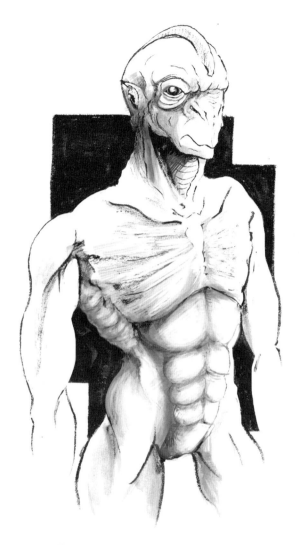

FIGURE 29.4 An intelligent extraterrestrial of the type beloved by science fiction. Of course, we have no idea what real intelligent extraterrestrials will look like. Indeed, they may take many different forms, given the trillions of exoplanets on which they probably exist.

named after Goldilocks's finding the third of three things (chairs, beds, bowls of porridge) to be 'just right'. The idea is that too close to a sun is a bad place for a planet to be in terms of its acting as a home for life; too far away is also a bad place to be; whereas somewhere in between is just right. It isn't possible to state a range of distances for this zone – also referred to as the Habitable Zone – because that depends on the size of the sun concerned. The bigger and hotter the sun, the further out is its Goldilocks Zone.

Given the above ball-park calculations, the likely number of planets in the Universe is so large that even if the *proportion* of them that lie in the relevant Goldilocks Zone is tiny, the *number* of those will still be enormous. So the probability of extraterrestrial life still looks very high, even when most planets have been excluded from consideration because of being in a place that is not 'just right'.

The clear conclusion of this chapter is that extraterrestrial life is a near-certainty, as is the existence of extraterrestrial animal-type life. Even extraterrestrial intelligent life seems nearly certain to exist because of the sheer scale of the Universe and the fact, only known from research over the last two or three decades, that the typical star (or sun) has one or more planets orbiting it.

Of course, *existence* of extraterrestrial intelligence and *discovery* of it are two very different things. As far as we know, nothing can travel faster than light. Radio waves, as part of the electromagnetic spectrum of radiation, travel at exactly the speed of light. Radio waves sent out into space by humans 50 years ago will by now have reached the nearest stars – but still only a small fraction of the stars in the Milky Way. Likewise, radio waves originating from planets orbiting nearby stars the same length of time ago would have reached us by now. But for more distant stars, such as Polaris, no contact would yet have been made on the basis of such broadcasts. It may be a while yet before we make the most incredible discovery of all – that indeed we are truly not alone.

At the start of this chapter, I made the point that extraterrestrial life need not be carbon-based or cellular, and need not use DNA as its genetic material – though equally, it may be characterized by all of those things. A related issue is whether extraterrestrial intelligence would be based on something like our own nervous system. Again, it might and it might not. In any event, for us the nervous system is the physical basis for our intelligence, and presumably for our consciousness too. So an examination of nerves, brains and consciousness seems an appropriate final chapter for this book.

30 The ghost in the machine

I have borrowed this phrase from the Hungarian-British writer Arthur Koestler, who in turn borrowed it – for use as a book title – from the English philosopher Gilbert Ryle, who devised it in the first place. Ryle used it as a derogatory term for criticizing the view that the mental and physical worlds are essentially separate. However, sometimes the meanings of clever phrases evolve from negative to neutral or even positive. This is true of 'the Big Bang', introduced by British astronomer Fred Hoyle to ridicule the idea of an expanding universe of finite age but now widely used by supporters of this view. Although an identical shift cannot be said to have happened in relation to 'the ghost in the machine', this phrase is often now used in a more neutral way than Ryle intended. I use it here because I like the phrase from a literary perspective; and it seems appropriate as a title for a chapter in which the main focus is animal consciousness. But I do not wish to prejudge at the outset the nature of the link between the brain and the consciousness of a human or any other kind of animal.

In previous chapters we have looked briefly at ideas about animal intelligence. For example, in Chapter 12 we considered whether the octopus might be the most intelligent of all invertebrates; and in Chapter 24 we examined the difference in intelligence between humans and our closest relatives, chimpanzees. Intelligence and consciousness are not the same thing, though they are related in the sense that both depend on the existence of a nervous system.

Or do they? From the standpoint of the last chapter we really can't be sure. So far, all our understanding of life comes from studies of the life-forms found on a single planet. So, although there have been many studies on nervous systems and consciousness, conducted on animals belonging to various species, our sample size is effectively just 'one'. And this, as any statistician will tell you, is not a good basis from which to make generalizations.

But perhaps, taking a more modest approach and attempting to generalize about life on Earth rather than life anywhere in the Universe, we can indeed be confident that the nervous system is a requirement for consciousness. Several observations point in this direction; here are three of them.

First, as far as we know, organisms from other kingdoms than the animal one are not conscious. This is true not only of simple unicellular creatures such as bacteria but also of impressively large and complex members of the plant kingdom, such as trees. Of course, there are eccentrics who disagree: we've all heard of tree-huggers, some of whom claim that the trees they hug are conscious. But, at least among the scientific community, it is hard to find anyone who will support the notion that trees are self-aware. By the way, note that I have not defined consciousness – it's notoriously hard to do so – but I have just given some indication of what I consider it to be by equating it with self-awareness.

Second, animals without nervous systems, the main group of which is the phylum Porifera (the sponges), appear not to be conscious. And animals *with* nervous systems, which normally seem to be conscious, lose their consciousness if their nervous system sustains sufficient damage. Regrettably, we have evidence of this from those humans who have ended up in what is referred to as a persistent vegetative state because of a serious accident. It's interesting to note that this term uses a reference to the plant kingdom; the implication is that, as argued above, plants are not conscious.

Third, it seems unlikely that humans are conscious as early embryos. Indeed, this supposition is among the arguments that underpin legislation, in many countries, dictating that abortion is acceptable before a certain time in embryogenesis but not after that time. If this is true, then consciousness is another feature to add to the list of previously discussed ones where we find that development in a sense recapitulates evolution. In early evolution, no organisms were conscious; consciousness took some time to evolve. In later evolution (the present, for example), those animals that are conscious are only so from a hard-to-define point that is some way into their developmental process. So the developmental sequence from non-conscious to conscious in these animals recapitulates the corresponding, but much longer-term, evolutionary sequence.

Let's now begin to refine our picture of the dependence of consciousness on a nervous system. I would say that the existence of a nervous system in an animal is a necessary but not sufficient condition for the animal concerned to be conscious. The rationale underlying this claim is

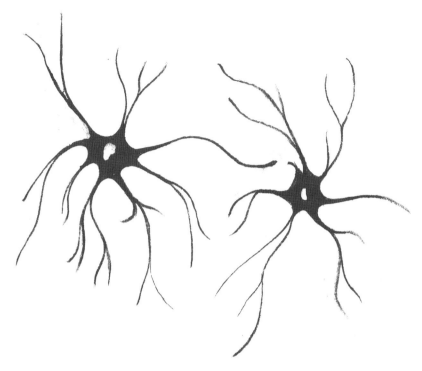

FIGURE 30.1 Nerve cells, also known as neurons. Both our central nervous system (brain plus spinal cord) and our peripheral nervous system are made up of neurons, of various kinds. Neurons interact with each other at synapses – places where the projections of one nerve cell contact those of another.

to be found in those animals that have nervous systems but rather rudimentary ones. For example, moving from sponges (no nervous system) to jellyfish (very simple nervous system), it is doubtful that we have moved from a non-conscious animal to a conscious one. A jellyfish has only a diffuse nerve network in its body; there is no large, dense concentration of nerves in any particular place. In other words, there is no brain. This line of argument suggests that not just a nervous system but a centralized one with a brain is a necessary (but still not sufficient?) prerequisite for consciousness.

At this point it seems appropriate to consider how nervous systems work. In a book of this kind, our examination of this topic will necessarily be brief; but a brief look at the subject is better than no look at all.

The basic unit of any nervous system is the nerve cell, otherwise known as a neuron (Figure 30.1). Even in simple nervous systems there

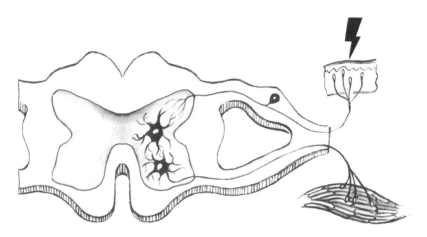

FIGURE 30.2 An example of a reflex reaction – the removal of a hand from a heat source on receiving a burn. Such reflex reactions do not involve the brain. In the reaction shown, a pain message received at the surface of the skin is relayed to a muscle via just three neurons. The short projections from neurons are called dendrites, the long ones axons.

are several types of neurons; and in more complex ones there are lots of types. A fundamental categorization is into sensory and motor neurons. The former, as their name implies, sense things – for example temperature, touch or pain. The latter conduct impulses that cause things to happen – for example, a muscle to contract. (There are also interneurons, which connect other neurons together.)

Even in animals with complex nervous systems, such as humans, simple interactions can occur involving just sensory neurons, interneurons and motor neurons, without the brain being involved in any way. A classic example of such a reflex reaction is the knee-jerk response to being tapped just below the kneecap. What happens here is that sensory neurons are activated by the tap and convey signals into the spinal cord. There, the axons of the sensory neurons connect with the dendrites of interneurons, which in turn pass on the signal to motor neurons. These convey the signals to the relevant muscles, which then contract, causing the leg below the knee to jerk forward. The retraction of a hand from a heat source on receiving a burn is a broadly similar kind of process (Figure 30.2).

In animals with simple nerve networks and no brain, all of the activity within the nervous system can be thought of as a series of reflex reactions. But in animals with brains many other types of neural activity are possible. To what extent this is true depends on the size and nature

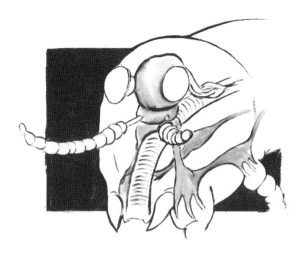

FIGURE 30.3 The head end of an insect, showing the concentration of nerve cells into groupings called ganglia (shaded). The largest (anterior-most) ganglion, or group of ganglia, can be called a brain. Here, the brain is shown at the top left. Lower down in the picture are smaller ganglia that innervate appendages.

of the brain. Indeed, although a brain is a physical thing, in contrast to consciousness, which is a mental one, it is not easy to define. Moving from sponges and jellyfish to the bilaterian animals, the ancestor of which was the urbilaterian discussed in Chapter 8, we find animals, including ourselves, which have brains that clearly deserve the title. However, we also find many other animals, such as insects and snails, where what is called the brain is just a smallish concentration of nerve cells at the anterior end.

Any aggregation of nerve cells at a particular place within the body can be referred to as a ganglion. But how large does a ganglion have to be – i.e. how many nerve cells does it have to be composed of – to be called a brain? There is no clear answer to this question. In insects, the head-end ganglion (or ganglia in the plural) may be little bigger than the segmental ganglia that are found further back in the body, in the rear part of the head, the thorax and the abdomen (Figure 30.3). Yet it is often referred to as a brain, whereas the segmental ganglia are not graced with this title. One reason for this situation is that there is not a single criterion – size – for being a brain. Rather, there are two, and the other is simply being 'up front' in the head.

Of course, if we accept this twin basis for the use of the term *brain*, then it follows that no non-bilaterally-symmetrical animals can be said to have a brain – because they do not have a head. And in fact this works. There are radially symmetrical animals which, unlike the jellyfish, had distant ancestors that did have brains. The most obvious example is the phylum Echinodermata – the starfish and their kin. These animals have only a diffuse nerve net. In general, heads and brains tend to go together. The fact that they can be lost as well as gained helps us to remember that evolution has no intrinsic tendency towards either bodily complexity in general or neural complexity in particular.

Let's focus, from here onwards, on head-end concentrations of nerve cells, which we'll call brains. Taking this approach, it's clear that brains can be very varied in size and in complexity of structure and function. The insect brain pictured in Figure 30.3 is near one end of such a spectrum. Mammalian brains, most notably those of primates and cetaceans (dolphins and whales) are at the other end. We generally associate consciousness with the mammalian end of the spectrum. However, since there is indeed a spectrum rather than a division into two or more discrete categories, it's difficult to draw a line and to say that all animals on the spectrum up to that point are non-conscious while all animals above that point are conscious.

What emerges from looking at things in this way is the inevitable conclusion that consciousness, like the size and complexity of the brains upon which it is based, is a spectrum of possibilities rather than a definite entity. There are degrees of consciousness in the animal kingdom.

This conclusion corresponds with everyday observations on a wide range of animals. Few people, if afforded the opportunity to intensively study the behaviour of chimps or dolphins, would claim that these magnificent animals are entirely lacking in consciousness. Going downward in terms of brain size and complexity to dogs, the idea that these have no consciousness at all does not ring true to me. However, I appreciate that there is not much of a scientific basis for this feeling. It is a feeling that is based only on personal intuition and on the knowledge that, while the brains of dogs are indeed smaller and less complex then those of chimps, they are nevertheless rather impressive structures (Figure 30.4).

Starting at the other end of the spectrum, I doubt that a fly is conscious in any meaningful sense of the term. And I would feel the same about a snail. However, I would not be so quick to rule out consciousness

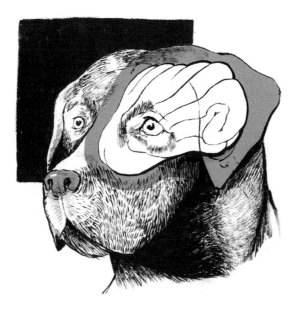

FIGURE 30.4 A dog's head, with its brain shown as if we could see inside. Despite the difficulty of defining intelligence, we can be reasonably confident that dogs are much more intelligent than insects. But the extent to which they are conscious remains an open question.

in the case of the snail's molluscan cousin, the octopus. This point re-emphasises the relatedness (but non-identity) of consciousness and intelligence.

Some biologists shy away from notions of consciousness because, although both it and intelligence are hard to define, the definitional problem with consciousness is more acute. The problem of making a link with patterns of behaviour is also more acute for consciousness – certain behaviour patterns are reasonably interpreted as implying a certain minimum level of intelligence, but it's harder to say that they also imply a minimum in terms of consciousness.

Other biologists, though, have been bolder and have attempted to get to grips with the issue of consciousness despite all the problems of defining and measuring it. However, it's not clear to what extent any have succeeded. The Maltese-British physician and writer Edward de Bono is probably most famous for his idea of lateral thinking. In 1969, he published the book *Mechanism of Mind*, which gives an excellent description of how the brain/mind works, using non-technical terms. His approach neatly captures many aspects of the way the human brain functions, including its self-organizing activity. But when it comes to explaining consciousness, de Bono's otherwise excellent model falls short. He suggests that all that is needed to explain a neural system becoming self-aware is that it develops a sort of mirror so that it can see itself. I wasn't convinced when I read his book as an undergraduate; and I'm still not convinced.

Perhaps the most famous biologist to have become involved in the study of consciousness was Francis Crick. About 15 years after publishing, with James Watson, the structure of DNA, Crick moved from Cambridge to the Salk Institute of Biological Studies in California. He switched his main research focus from molecular genetics to neuroscience and he focused in particular on finding a physical basis for consciousness. In 1994, he published *The Astonishing Hypothesis*, a popular exposition of this part of his life's work. Interestingly, he gave this book the subtitle *The Scientific Search for the Soul*.

Crick's subtitle raises an important, but some would say nonscientific, question. Do humans (and, for that matter, all 'higher animals') have a soul in the sense of something that can transcend their bodies and persist in some way after death? Although Crick thought not, one of the founders of the theory of natural selection – Alfred Russel Wallace – thought so; and it is interesting to examine his views.

Among Wallace's many books is the self-effacingly titled *Darwinism*, published in 1889. This can perhaps be thought of as Wallace's version of *On the Origin of Species*. There are chapters on the struggle for existence, variation, natural selection, difficulties with the theory, geographical distribution, the fossil record, and so on. The sequence of these is rather similar to the sequence of their equivalents in *The Origin*. The subtitle of Wallace's book is *An Exposition of the Theory of Natural Selection, with Some of its Applications*.

The final chapter in Wallace's book is "Darwinism applied to Man". This is where the two founders of the theory of natural selection differ most in their thinking. In previous chapters we have noted some other differences between Darwin and Wallace, including (in Chapter 27) Wallace's emphasis on the independent variation of different characters, in contrast to Darwin's emphasis on co-variation – what he called "correlation of growth". As we've seen, this is an important difference for evolutionary biologists because it has a bearing on the crucial question of whether the structure of variation may itself be an agent of evolutionary change ('yes' if we accept Darwin's view; 'no' if we accept Wallace's). It is perhaps not a difference that is of such great importance outside of biology, where it might be regarded as a technical matter (though see below for a counter-argument). But the difference between Darwin and Wallace on the origin of human consciousness and higher mental faculties is of broad philosophical significance. It's worth taking a closer look at this difference, as seen from Wallace's point of view at the time when he wrote *Darwinism*.

In the middle of his final chapter, Wallace has a section entitled "The origin of the moral and intellectual nature of Man". This section begins as follows:

> From the foregoing discussion it will be seen that I fully support
> Mr. Darwin's conclusion as to the essential identity of man's bodily
> structure with that of the higher mammalia, and his descent from some
> ancestral form common to man and the anthropoid apes. The evidence
> of such descent appears to me to be overwhelming and conclusive.

In his next paragraph, Wallace turns his attention to the bearing of Darwin's work on human "mental faculties" as opposed to "bodily structure":

> But this is only the beginning of Mr. Darwin's work, since he goes on
> to discuss the moral nature and mental faculties of man, and derives
> these too by gradual modification and development from the lower
> animals. Although, perhaps, nowhere distinctly formulated, his whole
> argument tends to the conclusion that man's entire nature and all his
> faculties, whether moral, intellectual, or spiritual, have been derived
> from their rudiments in the lower animals, in the same manner and by
> the same general laws as his physical structure has been derived.

Wallace then says that "this conclusion appears to me not to be supported by adequate evidence, and to be directly opposed to many well-ascertained facts".

There follows an explanation of this last point about "evidence" and "facts", which is not, in my view, persuasive. One of Wallace's main points is that he cannot believe that the higher mental faculties of humans, of which he singles out a few, such as mathematical ability, could have been produced through the action of natural selection. However, it is here that we see a connection between the different views of

Darwin and Wallace about the human mind and their different views on the independence versus co-variation of different characters – an argument that applies no less in relation to mental characters than in relation to physical ones. If, say, mathematical ability has never been the cause of fitness differences in human evolution, it may yet have spread under the influence of natural selection for some other character of the brain or mind that *was* such a cause, due to heritable character-correlation.

We have begun to stray away from consciousness. Let's return to it, first by looking at some more of Wallace's points in the concluding chapter of *Darwinism*. In fact, Wallace invokes three stages at which he claims that there were injections of something from the spirit world into the natural one. The first of these three involves the origin of life. Wallace argues that increases in chemical complexity in early biochemical evolution "could certainly not have produced *living* protoplasm". Instead, he argues that "we have indications of a new power at work, which we may term *vitality*". This makes it clear that Wallace was a vitalist at heart rather than a mechanist, a stance that does not sit comfortably with modern science. It also reveals a rather dogmatic approach: note the "could certainly not have produced" rather than a more guarded "seem unlikely to have been able to produce" (or equivalent).

Wallace's dogmatism continues when he gets specifically to the origin of consciousness, which is the second of the three stages at which he envisages some influence of the spirit world on the natural one having been exerted:

> The next stage is still more marvellous, still more completely beyond all possibility of explanation by matter, its laws and forces. It is the introduction of sensation or consciousness, constituting the fundamental distinction between the animal and vegetable kingdoms. Here all idea of mere complication of structure producing the result is out of the question.

Quite apart from any other criticism we might wish to make of this statement, the notion that all animals, including those without nervous systems, are conscious, is hard to accept.

Finally, and with continuing dogmatism:

> The third stage is, as we have seen, the existence in man of a number of his most characteristic and noblest faculties, those which raise him above

the brutes and open up possibilities of almost indefinite development. These faculties could not possibly have been developed by means of the same laws which have determined the progressive development of the organic world in general, and also of man's physical organism.

Although I am not a fan of the style of atheism espoused by Richard Dawkins in his 2006 book *The God Delusion*, I think that Wallace's views on the origin of life, consciousness and higher mental faculties reveal a slippery slope that we are in danger of sliding down if we start to accept even a single injection of something beyond the natural into arguments about life on Earth. Wallace does not stop at advocating one stage at which there was such an injection, but goes on to argue for the three noted above. And in fact Wallace prefaces his argument by stating that there are "at least three" such stages; so, although he was only explicit about three, he may have believed in more.

Where does this slippery slope end? Perhaps the origin of each of the animal phyla requires an injection of something from the spirit world? Maybe the origins of subphyla (such as the vertebrates) and classes also require a similar explanation? Perhaps even the origin of each species requires spiritual intervention, in contrast to the evolution of varieties within species, which can be explained in terms of natural processes? The answer to the question about the slippery slope is now clear: it ends, ironically, in some version or other of creationism. By allowing a single non-natural influence on the living world that is beyond scientific scrutiny, we enter into a domain of superstition. In my view, it is very important that this fate is avoided: it has the capacity to take human thinking back more than two centuries.

The views of Francis Crick on the nature of consciousness could hardly be more different than those of Wallace. In *The Astonishing Hypothesis*, Crick outlines a scientific approach to understanding consciousness, and what he calls "the soul" in his subtitle. His quest is for an explanation that is based on molecular biological processes and on the laws of physics and chemistry that underlie them. Never one to beat about the bush, Crick begins his book with a very clear statement of what his astonishing hypothesis is. It is "that 'You', your joys and your sorrows, your memories and your ambitions, your sense of personal identity and free will, are in fact no more than the behaviour of a vast assembly of nerve cells, and their associated molecules."

Most of Crick's book is based on a restricted approach to consciousness, centred on what might be called visual awareness. Crick takes the view that such awareness, and perhaps consciousness more generally, is based on two things: attention (e.g. to visual information) and short-term memory. However, he broadens out, at the end of the book, to discuss the issue of free will. He notes that studies of people with a particular kind of brain damage suggest that the main centre of free will in the brain is near an area called the anterior cingulate sulcus. In order to understand free will more comprehensively, Crick advocates not an extension of studies into the philosophical or religious realms but rather, in line with his overall approach, more experiments.

Crick's approach was based on the laws of what I'll call conventional science. There is, however, another approach that has been taken to the nature of consciousness that is scientific but rather unconventional. What I'm referring to here is the view taken by a small number of scientists that classical physics is inadequate to explain the mind and consciousness, because the mind works in a way that involves quantum effects. Foremost among these scientists are the British mathematical physicist Roger Penrose and the American anaesthesiologist Stuart Hameroff, who have argued specifically that quantum effects in cellular structures called microtubules are responsible for producing consciousness – see Penrose's 1994 book *Shadows of the Mind: A Search for the Missing Science of Consciousness* for an account. This, however, is a minority view within the scientific community.

Wallace's spiritual view of the nature of consciousness and the higher mental faculties of humans goes together with a belief in the persistence of the soul of each individual after death. In contrast, the most obvious interpretation of the soul in the approaches of Crick and of Penrose and Hameroff is that it is simply another word for the mind, and that when the brain dies the mind/soul is extinguished too. Most scientists, and indeed most people, believe one of these things or the other, or are undecided about which of them is true, perhaps oscillating in their view of the issue over the course of their lives. There is, however, another possibility. To see this, we will turn to one of the writings of the British geneticist J. B. S. Haldane, whose name came up previously (in Chapter 27) in relation to the important evolutionary issue of the structure of variation.

As well as writing various technical books and papers, Haldane wrote books belonging to popular science and to other, harder-to-classify genres. One of these is his 1927 collection of essays, *Possible Worlds*. One of the many and varied essays included therein is entitled "When I am dead". In this essay, Haldane spells out his view of the possibility of his mind's individual survival after death as "an improbable theory". However, he goes on to advance an interesting alternative, as follows. He says: "It seems to me immensely unlikely that mind is a mere by-product of matter." He then continues: "But as regards my own very finite and imperfect mind, I can see, by studying the effects on it of drugs, alcohol, disease, and so on, that its limitations are largely at least due to my body." Immediately following this he puts forward an interesting suggestion:

> Without that body it may perish altogether, but it seems to me quite as probable that it will lose its limitations and be merged into an infinite mind or something analogous to a mind which I have reason to believe probably exists behind nature. How this might be accomplished I have no idea.

I would love to know what was going on in Haldane's head when he wrote those lines. I am especially intrigued by his suggestion that the merging of his own mind into a universal mind on his death was equally probable to the extinction of his own mind. I am also intrigued by his use of the phrase "I have reason to believe". What reason, one wonders. There is a possibility that the whole essay was written tongue-in-cheek. A pointer in this direction is the footnote attached to the title of the essay. This reads: "This paper was one of a series on the above topic, most of which supported a different thesis." However, since several theses are advanced in the essay, this comment is ambiguous. Another possibility is that Haldane did believe in the "universal mind" idea, or at least consider it plausible, at the time of writing that particular essay. I doubt if historians of science will ever find out for certain.

The issue of the nature of the human mind/soul is linked with the issue of the existence or otherwise of a god, or of gods. However, this is not the place for an exploration of the debate between theists and atheists. Rather, in that whole domain, I suggest that we recall (from Chapter 1) the wise words written in 1886 by Darwin's bulldog, T. H. Huxley, coiner of the term *agnosticism*: "I am too much of a sceptic to deny the possibility of anything." Huxley's open-minded view has much to recommend it. We should not presume to know that science has all the answers – at least not yet. And the limitations of the science of the future can only be guessed at from our present-day perspective.

It is clear that much remains to be understood about the conscious-ness of humans and other animals here on Earth. Even more remains to be understood about the consciousness or otherwise of extraterrestrial life – in that right now we know nothing beyond the probability of existence of such life being very high. It's possible that the next decade or so will see our first radio contact with intelligent life-forms who inhabit an exoplanet orbiting a nearby star. The detection by humans of radio signals sent from another world will, if it happens, be one of the most momentous discoveries of our conscious, intelligent minds – minds that have been a long while in the making, through diverse evolutionary modifications of the nervous systems of our many animal ancestors, as they made their hazardous journeys through time.

Appendix

Throughout the book, I have adhered to a policy of keeping technical terms to a minimum. In particular, I have tried not to bury the reader in a mass of scientific names of animal groups and geological periods. I hope this has resulted in the main text being easier to read than it otherwise would have been. However, many readers will be curious to know how a particular geological period that has been referred to (e.g. Cambrian, Jurassic) fits into the bigger picture of geological time. Equally, many readers will want to see at a glance somewhere a list of the animal phyla that have been dealt with in this book (e.g. arthropods, chordates, molluscs), with a grouping of these that relates to their place in the hierarchical structure of the animal kingdom as a whole, and with an estimate of how many described species each currently has. Hence the following.

GEOLOGICAL TIME

The Earth is just over 4.5 billion years old. This span of time is divided up into four aeons: Hadean (4.5 to 4.0 billion years ago, BYA), Archaean (4.0 to 2.5 BYA), Proterozoic (2.5 to 0.542 BYA) and Phanerozoic (usually given in millions of years ago: 542 to 0 MYA). There are subdivisions of all four aeons, but we don't need to know these for the first three, with a single exception: the end of the Proterozoic aeon, from about 635 to 542 MYA, is called the Ediacaran period, after the Ediacara Hills in South Australia, the site of the first discoveries of fossils of what have come to be known as the Ediacaran life-forms (not 'animals', as that remains uncertain). So the creatures themselves, the place where they were first found (though it's now clear that they had a worldwide distribution) and the relevant period of geological time all bear the same name. This was not the case until quite recently, because 'Vendian period' was previously used for the approximate time span now called the Ediacaran.

When we come forward into the Phanerozoic aeon, it is more important to know about subdivisions of geological time. The primary subdivision is into three eras: Palaeozoic, Mesozoic and Cenozoic; these names translate as old, middle and young animals. The three eras start at 542, 250 and 65 MYA respectively. The eras, in turn, are divided into periods. The names of all these periods are given in Figure A1.

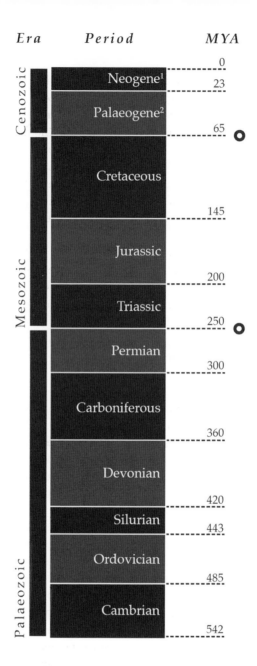

Era	Period	MYA

O *Mass Extinction Event*

FIGURE A1 The three most recent eras of geological time, together with their constituent periods. Most dates have been rounded to the nearest 5 million years. The mass extinctions at the end of the Palaeozoic and Mesozoic eras are indicated; other major extinction events are not. MYA stands for millions of years ago. Notes: (1) The last 2.5 million years form the Quaternary period, but there isn't sufficient room for this name to be shown in the figure. (2) The Tertiary was the old name for the span of time that is now split into the Palaeogene and Neogene periods.

The ends of the Palaeozoic and Mesozoic eras are defined by mass extinction events – the ones that occurred at 250 and 65 MYA. They are not the only mass extinction events in the history of the animal kingdom, but they are the biggest ones. Those iconic invertebrates the trilobites finally became extinct in the former event, while the even more iconic dinosaurs perished in the latter. However, it's always worth keeping in mind that in both events the majority of animal species disappeared, and that extinctions occurred across the board in terms of animal groups on both occasions.

The names of the various geological periods of the Phanerozoic era have remained quite stable over the last half century, with the exception of what used to be called the Tertiary period. This corresponds to what is now known as the Palaeogene and the Neogene (together spanning from 65 to about 2.5 MYA: see Figure A1). The word Tertiary is still often used though, as is the phrase K/T extinction, for the mass extinction event in which the dinosaurs died out. The T in this phrase stems from Tertiary, while the K relates to the Cretaceous. The latter initially seems illogical, but C is used for the Carboniferous period, so the Cretaceous is given a K (from *Kreide*, the German for chalk, a type of sedimentary rock that is characteristic of this period, especially in Europe).

The various geological periods are divided into epochs, and these can be further divided into ages. For the most part we don't need to go into that sort of detail. However, it's worth noting that some periods are divided into epochs that are simply called Upper (more recent) and Lower (older); or Upper, Middle and Lower. In other cases, the epochs are given their own names, for example we are currently in the Holocene epoch, the later of two epochs within the Quaternary period.

PHYLA OF THE ANIMAL KINGDOM

There are about 35 animal phyla. It's necessary to use 'about' because not all authors recognize the same phyla; this fact derives from the related one that there is no clear definition of a phylum. For example, most authors recognize the Arthropoda as a phylum; but some raise what others would call subphyla within the Arthropoda (Crustacea, etc.) to the phylum level.

In the hierarchical structure of the system of groups within groups that we call the animal kingdom, the phylum is the highest level (or rank) of group for which there is a taxon name. Thus, other named levels of taxa, in particular classes, orders, families, genera and species, are at lower levels in the hierarchy. However, there are names for *individual* groups above the phylum level; for example, we have come across all of the following in this

book: Bilateria, Protostomia, Deuterostomia, Ecdysozoa and Lophotrochozoa. I have generally referred to these as super-groups, to indicate that each
of them is a grouping of several phyla.

In the list that follows, I give the names of 15 phyla – essentially those
that have been discussed in this book. For the other 20 or so, I refer the
reader to the classic 2003 textbook *Invertebrates* by the Brusca brothers.
The list below groups phyla into the super-groups noted above, as well as a
'group' of basally branching phyla – this last 'group' is in inverted commas
because it is not a monophyletic one. Also, the list gives an indication of
common names, and it gives very approximate numbers of extant species.

> *Basally branching phyla*
>> Porifera (sponges, 8000)
>> Cnidaria (jellyfish, sea anemones, corals and others, 10,000)
>> Placozoa (flat animals, 1)
>
> *Deuterostomia*
>> Echinodermata (starfish, urchins, sea cucumbers, brittle stars and sea
>> lilies, 8000)
>> Hemichordata (acorn worms and pterobranchs, 100)
>> Chordata (vertebrates, cephalochordates and urochordates, 50,000)
>
> *Protostomia: Ecdysozoa*
>> Arthropoda (crustaceans, insects, arachnids and myriapods, about
>> 1.2 million)
>> Onychophora (velvet worms, 100)
>> Tardigrada (water-bears, 1000)
>> Nematoda (roundworms, 25,000)
>> Priapulida (penis-worms, 25)
>
> *Protostomia: Lophotrochozoa*
>> Mollusca (snails, slugs, bivalves, cephalopods and others, 100,000)
>> Annelida (segmented worms, 20,000)
>> Platyhelminthes (flatworms, 20,000)
>> Nemertea (ribbon worms, 1000)

Adding the approximate number of described extant species for each of the
named phyla gives between 1.4 and 1.5 million species in total. The 20 or so
phyla not included generally have small numbers of species, so the total
including them would probably still be in the region of 1.5 million. The
actual number of extant animal species (as opposed to the number of known
and named ones) is in excess of this by a hard-to-determine extent. Some
biologists think that the number of animal species still to be discovered may

be less than a million; others think that it may be as high as 30 million. The tropical rainforests and the deep oceans are recognized as places where most of the new discoveries will probably come from, though the soil ecosystem is another possible source of very large numbers of species that are still to be discovered.

In terms of number of known, extant species, the arthropods dwarf the rest of the animal kingdom. However, since by far the biggest arthropod group is the insects, which originated in the Silurian period, the arthropods may not have been as dominant in the early history of the animal kingdom, for example in the Cambrian, as they are today. However, they were a big group even then, including the trilobites and others (e.g. anomalocarids) that subsequently became extinct. As for the breakdown of the guesstimated 2–30 million species of animals that may actually exist at present, it's hard to be sure. There are probably very many arthropod and non-arthropod species yet to be discovered, with nematodes likely to provide many of the latter. And in relation to the animal kingdom of the distant evolutionary future, well, it's not even wise to attempt a guesstimate.

References

In line with the general ethos of the book, these references have informal pointers to them from the main text. It's clear when you encounter a reference-pointer, because each one has an author's name and a year in close proximity. Where a name occurs without a precise year, there is no corresponding entry in the reference list.

Aguinaldo, A. M. A., Turbeville, J. M., Linford, L. S., Rivera, M. C., Garey, J. R., Raff, R. A. & Lake, J. A. (1997) Evidence for a clade of nematodes, arthropods and other moulting animals. *Nature*, **387**, 489–493.

Allen, C. E., Beldade, P., Zwaan, B. J. & Brakefield, P. M. (2008) Differences in the selection response of serially repeated color pattern characters: Standing variation, development, and evolution. *BMC Evolutionary Biology*, **8**, doi:10.1186/1471-2148-8-94.

Arthur, W. (2004) *Biased Embryos and Evolution*. Cambridge University Press, Cambridge.

Asher, R. J. (2012) *Evolution and Belief: Confessions of a Religious Paleontologist*. Cambridge University Press, Cambridge.

Asher, R. J., Lin, K. H., Kardjilov, N. & Hautier, L. (2011) Variability and constraint in the mammalian vertebral column. *Journal of Evolutionary Biology*, **24**, 1080–1090.

Azevedo, R. B. R., Keightley, P. D., Lauren-Maatta, C., Vassilieva, L. L., Lynch, M. & Leroi, A. M. (2002) Spontaneous mutational variation for body size in *Caenorhabditis elegans*. *Genetics*, **162**, 755–765.

Bateson, W. (1894) *Materials for the Study of Variation, Treated with Especial Regard to Discontinuity in the Origin of Species*. Macmillan, London.

Beldade, P., Koops, K. & Brakefield, P. M. (2002) Developmental constraints versus flexibility in morphological evolution. *Nature*, **416**, 844–847.

Bonner, J. T. (1974) *On Development*. Harvard University Press, Cambridge, MA.

Brusca, R. C. & Brusca, G. J. (2003) *Invertebrates*, 2nd edition. Sinauer, Sunderland, MA.

Butterfield, N. J. (2000) *Bangiomorpha pubescens* n. gen., n. sp.: implications for the evolution of sex, multicellularity, and the Mesoproterozoic/Neoproterozoic radiation of eukaryotes. *Paleobiology*, **26**, 386–404.

Carroll, S. B., Grenier, J. K. & Weatherbee, S. D. (2005) *From DNA to Diversity: Molecular Genetics and the Evolution of Animal Design*, 2nd edition. Blackwell, Malden, MA.

Chipman, A. D. (2010) Parallel evolution of segmentation by co-option of ancestral gene regulatory networks. *BioEssays*, **32**, 60–70.

Clark, R. B. (1964) *Dynamics in Metazoan Evolution: The Origin of the Coelom and Segments*. Clarendon Press, Oxford.

Coates, M. I. & Clack, J. A. (1990) Polydactyly in the earliest known tetrapod limbs. *Nature*, **347**, 66–69.

Conway Morris, S. (1998) *The Crucible of Creation: The Burgess Shale and the Rise of Animals*. Oxford University Press, Oxford.

Couso, J. P. (2009) Segmentation, metamerism and the Cambrian explosion. *International Journal of Developmental Biology*, **53**, 1305–1316.

Crick, F. H. C. (1994) *The Astonishing Hypothesis: The Scientific Search for the Soul*. Scribners, New York.

Darwin, C. (1859) *On the Origin of Species by Means of Natural Selection, or the Preservation of Favoured Races in the Struggle for Life*. J. Murray, London.

Dawkins, R. (1986) *The Blind Watchmaker*. Longman, London.

Dawkins, R. (2006) *The God Delusion*. Bantam Press, London.

de Bono, E. (1969) *The Mechanism of Mind*. Cape, London.

Duboule, D. (1994) Temporal collinearity and the phylotypic progression: a basis for the stability of a vertebrate Bauplan and the evolution of morphologies through heterochrony. *The Evolution of Developmental Mechanisms (Development 1994 Supplement)*. Company of Biologists, Cambridge.

Dugon, M. & Arthur, W. (2012) Comparative studies on the structure and development of the venom-delivery system of centipedes, and a hypothesis on the origin of this evolutionary novelty. *Evolution & Development*, **14**, 128–137.

Erwin, D. H. (2000) Macroevolution is more than repeated rounds of microevolution. *Evolution & Development*, **2**, 78–84.

Fabricius, H. (1621). De Formatione Ovi et Pulli. Padova. Translation in Adelmann, H. B. (1942) *The Embryological Treatises of Hieronymus Fabricius of Acquapendente*. Cornell University Press, New York.

Finn, J. K., Tregenza, T. & Norman, M. D. (2009) Defensive tool use in a coconut-carrying octopus. *Current Biology*, **19**, R1069–R1070.

Fisher, R. A. (1918) The correlations between relatives on the supposition of Mendelian inheritance. *Transactions of the Royal Society of Edinburgh*, **52**, 399–433.

Fortey, R. (2009) *Fossils: The Key to the Past*. Natural History Museum, London.

Galis, F., Van Dooren, T. J. M., Feuth, J. D., Metz, J. A. J., Witkam, A., Rulnard, S., Stelgenga, M. J. & Wijnaendts, L. C. D. (2006) Extreme selection in humans against homeotic transformations of cervical vertebrae. *Evolution*, **60**, 2643–2654.

Geoffroy Saint-Hilaire, E. (1822) Considérations générales sur la vertèbre. *Mémoires du Muséum d'Histoire Naturelle*, **9**, 89–119.

Gilbert, S., Loredo, G. A., Brukman, A. & Burke, A. C. (2001) Morphogenesis of the turtle shell: the development of a novel structure in tetrapod evolution. *Evolution & Development*, **3**, 47–58.

Goldschmidt, R. (1952) Homeotic mutants and evolution. *Acta Biotheoretica*, **10**, 87–104.

Gould, S. J. (1977) *Ontogeny and Phylogeny*. Harvard University Press, Cambridge, MA.

Gould, S. J. (1989) *Wonderful Life: The Burgess Shale and the Nature of History.* Hutchinson Radius, London.

Gould, S. J. (1993) *Eight Little Piggies: Reflections in Natural History.* Jonathan Cape, London.

Gould, S. J. (1995) *Dinosaur in a Haystack: Reflections in Natural History.* Harmony, New York.

Gould, S. J. & Lewontin, R. C. (1979) The spandrels of San Marco and the Panglossian paradigm: a critique of the adaptationist programme. *Proceedings of the Royal Society of London, B,* **205**, 581–598.

Grant, P. R. & Grant, B. R. (2008) *How and Why Species Multiply: The Radiation of Darwin's Finches.* Princeton University Press, Princeton, NJ.

Gurdon, J. B., Elsdale, T. R. & Fischberg, M. (1958) Sexually mature individuals of *Xenopus laevis* from the transplantation of single somatic nuclei. *Nature,* **182**, 64–65.

Haeckel, E. (1866) *Generelle Morphologie der Organismen.* Georg Reimer, Berlin.

Haeckel, E. (1896) *The Evolution of Man : A Popular Exposition of the Principal Points of Human Ontogeny and Phylogeny.* Appleton, New York.

Haeckel, E. (1900) *The Riddle of the Universe at the End of the Nineteenth Century.* Watts, London.

Haldane, J. B. S. (1927) *Possible Worlds and Other Essays.* Chatto and Windus, London.

Haldane, J. B. S. (1932) *The Causes of Evolution.* Longman, London.

Hedges, S. B. & Kumar, S. (eds) (2009) *The Timetree of Life.* Oxford University Press, Oxford.

Hejnol, A. & Martindale, M. Q. (2009) The mouth, the anus, and the blastopore – open questions about questionable openings. In M. J. Telford & D. T. J. Littlewood (eds) *Animal Evolution: Genomes, Fossils, and Trees,* pp. 33–40. Oxford University Press, Oxford.

Hennig, W. (1966) *Phylogenetic Systematics.* University of Illinois Press, Urbana.

Huxley, T.H. (1886). Letter to Herbert Spencer. Published in L. Huxley (1903) *The Life and Letters of Thomas Henry Huxley,* vol 2, p443.

Huxley, T.H. (1894). Biogenesis and Abiogenesis. In *Collected Essays,* vol. 2, p. 244. Macmillan, London.

Jacob, F. (1977) Evolution and tinkering. *Science,* **196**, 1161–1166.

Kaji, T., Moller, O. S. & Tsukagoshi, A. (2011) A bridge between original and novel states: ontogeny and function of "suction discs" in the Branchiura (Crustacea). *Evolution & Development,* **13**, 119–126.

King, N., Westbrook, M. J., Young, S. L. et al. (2008) The genome of the choanoflagellate *Monosiga brevicolis* and the origin of metazoans. *Nature,* **451**, 783–788.

Leroi, A. M. (2000) The scale independence of evolution. *Evolution & Development,* **2**, 67–77.

Lewontin, R. C. (1974) *The Genetic Basis of Evolutionary Change.* Columbia University Press, New York.

Linnaeus, C. (1735) *Systema Naturae sive Regna Tria Naturae.* Haak, Leiden.

McGhee, G. (2011) *Convergent Evolution: Limited Forms Most Beautiful.* MIT Press, Cambridge, MA.

McGinnis, W., Garber, R. L., Wirz, J., Kuroiwa, A. & Gehring, W. J. (1984) A homologous protein-coding sequence in *Drosophila* homeotic genes and its conservation in other metazoans. *Cell,* **37**, 403–408.

McMenamin, M. A. S. (1998) *The Garden of Ediacara: Discovering the First Complex Life.* Columbia University Press, New York.

Mendel, G. (1866) Versuche über Pflanzenhybriden. *Verhandlungen des natur-forschenden Vereines in Brünn, Bd. IV für das Jahr 1865, Abhandlungen,* 3–47.

Minelli, A., Boxshall, G. & Fusco, G. (eds) (2013) *Arthropod Biology and Evolution: Molecules, Development, Morphology.* Springer, Berlin.

Nielsen, C. (2012) *Animal Evolution: Interrelationships of the Living Phyla,* 3rd edition. Oxford University Press, Oxford.

Nüsslein-Volhard, C. & Wieschaus, E. (1980) Mutations affecting segment number and polarity in *Drosophila. Nature,* **287**, 795–801.

Penrose, R. (1994) *Shadows of the Mind: A Search for the Missing Science of Consciousness.* Oxford University Press, Oxford.

Pough, F. H., Janis, C. M. & Heiser, J. B. (2002) *Vertebrate Life,* 6th edition. Prentice Hall, Upper Saddle River, NJ.

Raff, R. A. & Kaufman, T. C. (1983) *Embryos, Genes and Evolution: The Developmental Genetic Basis of Evolutionary Change.* Macmillan, New York.

Richardson, M. K. (1995) Heterochrony and the phylotypic period. *Developmental Biology,* **172**, 412–421.

Robert, J. S. (2004) *Embryology, Epigenesis and Evolution: Taking Development Seriously.* Cambridge University Press, Cambridge.

Sander, K. (1983) The evolution of patterning mechanisms: gleanings from insect embryogenesis and spermatogenesis. In B. C. Goodwin, N. Holder & C. C. Wylie (eds) *Development and Evolution,* pp. 137–159. Cambridge University Press, Cambridge.

Scott, M. P. & Weiner, A. J. (1984) Structural relationships among genes that control development: sequence homology between the *Antennapedia, Ultrabithorax* and *fushi tarazu* loci of *Drosophila. Proceedings of the National Academy of Sciences of the USA,* **81**, 4115–4119.

Shubin, N. (2008) *Your Inner Fish: A Journey into the 3.5-Billion-Year History of the Human Body.* Vintage, New York.

Simpson, G. G. (1953) *The Major Features of Evolution.* Columbia University Press, New York.

Smith, J. L. B. (1939) A living fish of Mesozoic type. *Nature,* **143**, 455–456.

Taylor, S. R. (2012) *Destiny or Chance Revisited: Planets and their Place in the Cosmos.* Cambridge University Press, Cambridge.

Telford, M. J. & Littlewood, D. T. J. (eds) (2009) *Animal Evolution: Genomes, Fossils and Trees.* Oxford University Press, Oxford.

Thompson, D'A. W. (1917) *On Growth and Form.* Cambridge University Press, Cambridge.

Thomson, K. S. (1991) *Living Fossil: The Story of the Coelacanth*. Hutchinson Radius, London.

von Baer, K. E. (1828) *Über Entwicklungsgeschichte der Tiere: Beobachtung und Reflexion*. Bornträger, Königsberg.

Waddington, C. H. (1956) Genetic assimilation of the bithorax phenotype. *Evolution*, **10**, 1–13.

Wallace, A. R. (1870) *Contributions to the Theory of Natural Selection: A Series of Essays*. Macmillan, London.

Wallace, A. R. (1889) *Darwinism: An Exposition of the Theory of Natural Selection, with Some of its Applications*. Macmillan, London.

Watson, J. D. & Crick, F. H. C. (1953) Molecular structure of nucleic acids: a structure for deoxyribose nucleic acid. *Nature*, **171**, 737–738.

Wharton, D. A. (2002) *Life at the Limits: Organisms in Extreme Environments*. Cambridge University Press, Cambridge.

Whyte, L. L. (1965) *Internal Factors in Evolution*. Tavistock Publications, London.

Wray, G. A., Levinton, J. S. & Shapiro, L. H. (1996) Molecular evidence for deep pre-Cambrian divergences among metazoan phyla. *Science*, **274**, 568–573.

Index